Builders' Estimating Data

Builders' Estimating Data

David M. G. Cross

Heinemann Newnes

Heinemann Newnes
An imprint of Heinemann Professional Publishing Ltd
Halley Court, Jordan Hill, Oxford OX2 8EJ

OXFORD LONDON MELBOURNE AUCKLAND SINGAPORE
IBADAN NAIROBI GABORONE KINGSTON

First published 1990

© David M. G. Cross 1990

British Library Cataloguing in Publication Data
Builders' estimating data.
 1. Construction industries. Costing
 I. Title
   624.0681

ISBN 0 434 90250 0

Typeset by Electronic Village Ltd, Richmond, Surrey
Printed by Butler & Tanner Ltd, Frome and London

Contents

Introduction

This book has been written for the practising building estimator in a small or medium-sized contracting firm. It should also be of interest to those personnel involved in the ordering, purchase and control of materials for building contracts.

It has been designed to aid the quick, accurate and consistent production of building estimates, using the minimum amount of calculation. By the use of pre-calculated constants when building up an estimate, the time taken to calculate the cost of labour, plant and materials required is greatly reduced. This method also eliminates the need for the estimator to keep detailed notes on how a particular unit rate was calculated.

However, the type, size and nature of building contracts, and of contractors, vary greatly: it is impossible to produce constants that will apply to every type of contract and contractor. The object of this book is to establish a *realistic starting point* from which estimators can adjust rates to suit their own requirements.

The material constants shown in the book include for nominal wastage, so any excessive waste of material can also be checked.

How to use this book

To calculate unit rates, simply multiply the cost of the labour, plant and materials by the constants given.

Example: Brick Wall

To calculate a unit rate for a half brick thick wall built in common bricks, laid stretcher bond in cement mortar 1:3 mix: first turn to Section F – Masonry, and calculate the cost of the mortar per cubic metre using the constants given in Section FA – Mortar Mixes.

O.P. cement	£65.00 per tonne × 0.470 =	30.55
Building sand	£6.35 per tonne × 1.780 =	11.30
Admix	£0.50 per litre × 2.350 =	1.18
5/3.5 mixer	£0.50 per hour × 1.330 =	0.67 = £43.70 per cu. m

Next turn to Section F10 – Brick and Block Walling, item F10.005, Common bricks, 215 × 102.5 × 65 mm in any mortar, and item F10.010, Walls, 102.5 mm, stretcher bond:

Common bricks	£0.10 per each × 60.440 =	6.04
Mortar (1:3)	£43.70 per cu. m × 0.032 =	1.40
2 + 1 Bricklaying gang	£4.16 per hour × 1.480 =	6.16 = £13.60 per sq. m

The above prices are nett. Overheads and profit can be added to the rates or at the end of the estimate as a lump sum.

Labour	=	£6.16 + 55% =	9.55
Materials and plant	=	£7.44 + 20% =	8.93 = £18.48 per sq. m. Unit Rate

Example: Plastering

To calculate a unit rate for 13 mm thick Carlite pre-mixed plaster to walls: first turn to Section M – Surface Finishes. Next turn to Section M20 – Plastered Coatings, item M20.205, Plaster, Carlite; pre-mixed; floating coat of browning, 11 mm thick; finish coat of finish, 2 mm thick; steel trowelled; internal; and item M20.210 13, Work to walls on brickwork or blockwork base; over 300 mm wide.

Browning	£62.00 per tonne × 0.0071 =	0.43
Finish plaster	£73.60 per tonne × 0.0022 =	1.40
2 + 1 Plastering gang	£4.16 per hour × 0.4600 =	1.91 = £3.74 per sq.m

Labour	=	£1.91 + 55% =	2.96
Materials and plant	=	£1.83 + 20% =	2.20 = £5.16 per sq. m. Unit Rate

Example: Drain Trench Excavation

To calculate a unit rate for excavating trenches (by machine) to receive pipes not exceeding 200 mm nominal size; grading bottoms; earthwork support; disposing of surplus excavated material by removing from site: first tune to Section D – Groundwork, item D20.505, Disposal of excavated material, and item D20.520, Removing from site to tip and paying tipping charges, average distance from site 15 km, using the constants given to calculate the cost per cu. m for cart away.

Machine loading	£12.00 per hour × 0.070 = 0.84
Wagon transporting	£15.00 per hour × 0.250 = 3.75
Tipping charges	£1.20 per cu. m × 1.000 = 1.20 = £5.79 per cu. m

Next turn to Section 12 – Drainage Below Ground, item R12.005, Excavating trenches to receive pipes, and item R12.010, Excavating; filling in above 400 mm thick beds or coverings with material arising from excavation not exceeding 2.0 m deep; average depth 500 mm.

Labour hours	£3.76 per hour × 0.080 = 0.30
Machine excavation	£12.00 per hour × 0.080 = 0.96
Plant compactor	£1.20 per hour × 0.010 = 0.01
Cart away	£5.79 per cu. m × 0.180 = 1.04 = £2.31 per ln. m

| Labour | £0.30 + 55% = 0.47 |
| Materials and plant | £1.83 + 20% = 2.01 = £2.48 per ln. m. Unit Rate |

Calculation of hourly labour rates

1. Annual working hours

Working weeks	52 weeks	30 weeks – Summer-time	
		22 weeks – Winter-time	
Public holidays	8 days Good Friday Easter Monday May Day Holiday Spring Bank Holiday Summer Bank Holiday	5 days – Summer-time	
	Christmas Day Boxing Day New Year's Day	3 days – Winter-time	
Annual holidays	21 days	14 days – Summer-time	
		7 days – Winter-time	

Non-productive overtime Say 1 hour per day,
5 days per week – Summer-time

30 weeks $\times$ 5 days $\times$ 1 hour	= 150.00 hours
Less:	
Annual holidays = 14 days $\times$ 1 hour	= 14.00 hours
Public holidays = 5 days $\times$ 1 hour	= 5.00 hours
	131.00 hours
131 hours $\times$ 1.50	= 196.50 hours
131 hours $\times$ 1.00	= 131.00 hours
Non-productive overtime	= 65.50 hours

2. Calculation of annual working hours

Summer-time working, 30 weeks at 44 hours per week, Monday to Friday		–	1320.00	

Less: Annual holidays – 14 days	123.00			
Public holidays – 5 days	44.00	167.20	1152.80

Winter-time working, 22 weeks at 39 hours per week,
Monday to Friday   –   858.00

Less: Annual holidays – 7 days   54.60
      Public holidays – 3 days   23.40
      Sick leave – 5 days   39.00   117.00   741.00

Total number of paid working hours per annum   1893.80
Less: Allowance for inclement weather, say 2%   37.80

Total number of productive hours worked per annum   1856.00

3. Calculation of labour costs   **Craftsman**   **Labourer**

Basic rate	1893.80 hours at	=		at	=	
Non-productive overtime	65.50 hours at	=		at	=	
Public holidays	63.00 hours	=		at	=	
Sick pay	5.00 days	=		at	=	
Tool money	45.20 weeks	=		at	=	
Plus rate	2022.30 hours	=		at	=	

Total cost of 1856 productive hours

Craftsman = _____ divide by 1856 productive hours = _____ Nett Hourly Rate

Labourer = _____ divide by 1856 productive hours = _____ Nett Hourly Rate

Note: to calculate an hourly rate for a two-and-one gang, add the hourly rates for two craftsmen and one labourer and divide by three.

D GROUNDWORK

DA

Notes

I   The labour constants in this section are based on one labourer.

2   The machine excavation and wagon transport hour constants include drivers' time, therefore the cost per hour of the machine and the wagon should include the cost of the driver.

3   The constants for plant compacting and breaking hours are for the item of plant only, the operators' time is included in the labour constant.

4   Where a constant is given for cart away per cu. m, the cost of carting away should first be calculated using the constants given in Section D20 – Disposal of Excavated Material, then multiplied by the constant given.

D20 *Excavating and filling*

D20.005	**Site preparation**
D20.010	Removing trees and grubbing up roots; removing from site; 600–1500 mm girth
D20.015	Removing trees and grubbing up roots; removing from site; 1500–3000 mm girth
D20.020	Removing trees and grubbing up roots; removing from site; over 3000 mm girth
D20.025	Clearing site of bushes, scrub, undergrowth, small trees and grubbing up roots
D20.030	Lifting turf to be preserved and stacking on site
D20.105	**Excavating by machine (JCB)**
D20.110	Topsoil for preservation
D20.115	To reduce levels
D20.120	Basements and the like
D20.125	Pits
D20.130	Trenches, width exceeding 0.30 m
D20.135	Trenches, width not exceeding 0.30 m
D20.140	For pile caps and ground beams between piles
D20.150	**Items extra over any type of excavation**
D20.155	Excavating below ground water level
D20.160	Next to existing services
D20.165	Around existing services crossing excavation

D20.170	**Breaking out existing materials**
D20.175	Rock
D20.180	Plain concrete
D20.185	Reinforced concrete
D20.190	Brickwork, blockwork or stonework
D20.195	Coated macadam or asphalt
D20.205	**Working space allowance, filling with excavated material**
D20.210	Reduce levels, basements and the like
D20.215	Pits
D20.220	Trenches
D20.225	Pile caps and ground beams between piles
D20.230	**Working space allowance, filling with imported hardcore**
D20.235	Reduce levels, basements and the like
D20.240	Pits
D20.245	Trenches
D20.250	Pile caps and ground beams between piles

D20.305	**Excavating by hand**
D20.310	Topsoil for preservation
D20.315	To reduce levels
D20.320	Basements and the like
D20.325	Pits
D20.330	Trenches, exceeding 0.30 m in width,
D20.335	Trenches, not exceeding 0.30 m in width,
D20.340	Pile caps and ground beams between piles
D20.350	**Items extra over any type of excavation**
D20.355	Below ground water level
D20.360	Next to existing services
D20.365	Around existing services crossing excavation
D20.370	**Breaking out existing materials**
D20.375	Rock
D20.380	Plain concrete
D20.385	Reinforced concrete
D20.390	Brickwork, blockwork or stonework
D20.395	Coated macadam or asphalt

D20.405	**Working space allowance, filling with excavated material**
D20.410	Reduce levels, basements and the like
D20.415	Pits
D20.420	Trenches
D20.425	Pile caps and ground beams between piles
D20.450	**Working space allowance, filling with imported hardcore**
D20.455	Reduce levels, basements and the like
D20.460	Pits
D20.465	Trenches
D20.470	Pile caps and ground beams between piles
D20.505	**Disposal of excavated material (machine loading – JCB)**
D20.510	Depositing on site in temporary spoil heaps
D20.515	Depositing on site, spreading and levelling
D20.520	Removing from site to tip and paying tipping charges
D20.525	Loading, transporting and depositing in spoil heaps
D20.550	**Disposal of excavated material (hand loading)**
D20.555	Depositing on site in temporary spoil heaps
D20.560	Depositing, spreading and levelling
D20.565	Removing from site to tip and paying tipping charges
D20.570	Loading, transporting and depositing in spoil heaps

D20   EXCAVATING AND FILLING

	Labour (hr)	Machine excavating (hr)	Wagon transport (hr)	Hardcore (tonne)	Concrete (cu. m)	Unit
D20.005	**Site preparation**					
D20.010	Removing trees and grubbing up roots; removing from site 600–1500 mm girth	12.00	1.25	0.50	—	—	no.
	Extra for filling holes with						
	excavated material	0.20	0.10	—	—	—	no.
	hardcore	0.40	0.20	—	2.03	—	no.
	lean mix concrete	3.66	—	—	—	1.00	no.
D20.015	Removing trees and grubbing up roots; removing from site 1500–3000 mm girth	45.00	3.00	1.75	—	—	no.
	Extra for filling holes with						
	excavated material	0.45	0.23	—	—	—	no.
	hardcore	0.91	0.45	—	4.56	—	no.
	lean mix concrete	8.24	—	—	—	2.25	no.
D20.020	Removing trees and grubbing up roots, removing from site over 3000 mm girth	70.00	3.50	2.75	—	—	no.
	Extra for filling holes with						
	excavated material	0.51	0.25	—	—	—	no.
	hardcore	1.01	0.51	—	5.06	—	no.
	lean mix concrete	9.15	—	—	—	2.50	no.
D20.025	Clearing site of bushes, scrub, undergrowth and grubbing up roots; cutting down small trees, not exceeding 600 mm girth	0.01	0.01	—	—	—	sq. m
D20.030	Lifting turf to be preserved and stacking on site	0.30	—	—	—	—	sq. m

	Labour (hr)	Machine excavating (hr)				Unit
D20.105	**Excavating by machine (JCB)**					
D20.110	Topsoil for preservation, average depth					
150 mm	0.01	0.01	—	—	—	sq. m
300 mm	0.02	0.02	—	—	—	sq. m
D20.115	To reduce level, maximum depth not exceeding					
0.25 m	0.15	0.15	—	—	—	cu. m
1.00 m	0.12	0.12	—	—	—	cu. m
2.00 m	0.14	0.14	—	—	—	cu. m
4.00 m	0.14	0.14	—	—	—	cu. m
D20.120	Basements and the like, maximum depth not exceeding					
0.25 m	0.15	0.15	—	—	—	cu. m
1.00 m	0.11	0.11	—	—	—	cu. m
2.00 m	0.13	0.13	—	—	—	cu. m
4.00 m	0.13	0.13	—	—	—	cu. m
D20.125	Pits, maximum depth not exceeding					
0.25 m	0.33	0.33	—	—	—	cu. m
1.00 m	0.22	0.22	—	—	—	cu. m
2.00 m	0.22	0.22	—	—	—	cu. m
4.00 m	0.27	0.27	—	—	—	cu. m
D20.130	Trenches, width exceeding 0.30 m, maximum depth not exceeding					
0.25 m	0.40	0.40	—	—	—	cu. m
1.00 m	0.33	0.33	—	—	—	cu. m
2.00 m	0.36	0.36	—	—	—	cu. m
4.00 m	0.40	0.40	—	—	—	cu. m
D20.135	Trenches, width not exceeding 0.30 m, maximum depth not exceeding					
0.25 m	0.42	0.42	—	—	—	cu. m
1.00 m	0.35	0.35	—	—	—	cu. m
2.00 m	0.38	0.38	—	—	—	cu. m
4.00 m	0.42	0.42	—	—	—	cu. m
D20.140	For pile caps and ground beams between piles, maximum depth not exceeding					
0.25 m	3.50	0.20	—	—	—	cu. m
1.00 m	3.75	0.20	—	—	—	cu. m
2.00 m	4.25	0.22	—	—	—	cu. m

		Labour (hr)	Machine excavating (hr)	Plant breaking (hr)	Hardcore (tonne)	Cart away (cu. m)	Unit
D20.150	**Items extra over any type of excavation**						
D20.155	Excavating below ground water level	0.50	0.50	—	—	—	cu. m
D20.160	Next to existing services	0.17	0.17	—	—	—	ln. m
D20.165	Around existing services crossing excavation	0.33	0.33	—	—	—	no.
D20.170	**Breaking out existing materials**						
D20.175	Rock	0.66	0.66	0.66	—	—	cu. m
D20.180	Plain concrete	0.33	0.33	0.33	—	—	cu. m
D20.185	Reinforced concrete						
	100 mm thick	0.05	0.05	0.05	—	—	sq. m
	150 mm thick	0.10	0.10	0.10	—	—	sq. m
	225 mm thick	0.15	0.15	0.15	—	—	sq. m
D20.190	Brickwork, blockwork or stonework						
	100 mm thick	0.05	0.05	0.05	—	—	sq. m
	150 mm thick	0.08	0.08	0.08	—	—	sq. m
	225 mm thick	0.11	0.11	0.11	—	—	sq. m
	338 mm thick	0.17	0.17	0.17	—	—	sq. m
D20.195	Coated macadam or asphalt						
	50 mm thick	0.01	0.01	0.01	—	—	sq. m
	100 mm thick	0.03	0.03	0.03	—	—	sq. m

		Labour (hr)	Machine excavating (hr)	Plant compacting (hr)	Hardcore (tonne)	Cart away (cu. m)	Unit
D20.205	**Working space allowance to excavations, filling with excavated material, compacting with a whacker in 250 mm layers**						
D20.210	Reduce levels, basements and the like	0.21	0.15	0.12	—	—	sq. m
D20.215	Pits	0.32	0.26	0.12	—	—	sq. m
D20.220	Trenches	0.36	0.30	0.12	—	—	sq. m
D20.225	Pile caps and ground beams between piles	0.72	0.15	0.12	—	—	sq. m
D20.230	**Working space allowance to excavations, filling with imported hardcore, compacting with a whacker in 250 mm layers, disposing of surplus excavated material by removing from site**						
D20.235	Reduce levels, basements and the like	0.31	0.19	0.12	1.215	0.60	sq. m
D20.240	Pits	0.37	0.25	0.12	1.215	0.60	sq. m
D20.245	Trenches	0.44	0.32	0.12	1.215	0.60	sq. m
D20.250	Pile caps and ground beams between piles	0.88	0.16	0.12	1.215	0.60	sq. m

	Labour (hr)					Unit
D20.305	**Excavating by hand**					
D20.310	Topsoil for preservation, average depth					
	150 mm	0.30	—	—	—	—	sq. m
	300 mm	0.60	—	—	—	—	sq. m
D20.315	To reduce levels, maximum depth not exceeding					
	0.25 m	2.85	—	—	—	—	cu. m
	1.00 m	3.00	—	—	—	—	cu. m
	2.00 m	3.75	—	—	—	—	cu. m
	4.00 m	6.00	—	—	—	—	cu. m
D20.320	Basements and the like, maximum depth not exceeding					
	0.25 m	2.85	—	—	—	—	cu. m
	1.00 m	3.00	—	—	—	—	cu. m
	2.00 m	3.75	—	—	—	—	cu. m
	4.00 m	6.00	—	—	—	—	cu. m
D20.325	Pits, maximum depth not exceeding					
	0.25 m	3.14	—	—	—	—	cu. m
	1.00 m	3.30	—	—	—	—	cu. m
	2.00 m	4.13	—	—	—	—	cu. m
	4.00 m	6.60	—	—	—	—	cu. m
D20.330	Trenches, exceeding 0.30 m in width, maximum depth not exceeding					
	0.25 m	3.14	—	—	—	—	cu. m
	1.00 m	3.30	—	—	—	—	cu. m
	2.00 m	4.13	—	—	—	—	cu. m
	4.00 m	6.60	—	—	—	—	cu. m
D20.335	Trenches, not exceeding 0.30 m in width, maximum depth not exceeding					
	0.25 m	0.24	—	—	—	—	ln. m
	0.50 m	0.50	—	—	—	—	ln. m
	0.75 m	0.74	—	—	—	—	ln. m
	1.00 m	0.99	—	—	—	—	ln. m
D20.340	Pile caps and ground beams between piles, maximum depth not exceeding					
	0.25 m	4.00	—	—	—	—	cu. m
	1.00 m	4.40	—	—	—	—	cu. m
	2.00 m	5.50	—	—	—	—	cu. m

		Labour (hr)	Plant compacting/ breaking (hr)	Hardcore (tonne)	Cart away (cu. m)	Unit
D20.350	**Items extra over any type of excavation**					
D20.355	Below ground water level	3.30	—	—	—	—	cu. m
D20.360	Next to existing services	0.40	—	—	—	—	ln. m
D20.365	Around existing services crossing excavation	0.66	—	—	—	—	no.
D20.370	**Breaking out existing materials**					
D20.375	Rock					
D20.380	Plain concrete	6.67	—	3.50	—	—	cu. m
D20.385	Reinforced concrete					
	100 mm thick	1.00	—	0.53	—	—	sq. m
	150 mm thick	1.50	—	0.80	—	—	sq. m
	225 mm thick	2.00	—	1.06	—	—	sq. m
D20.390	Brickwork, blockwork or stonework					
	100 mm thick	0.88	—	0.30	—	—	sq. m
	150 mm thick	1.31	—	0.45	—	—	sq. m
	225 mm thick	1.97	—	0.68	—	—	sq. m
D20.395	Coated macadam or asphalt					
	50 mm thick	0.17	—	0.09	—	—	sq. m
	100 mm thick	0.33	—	0.18	—	—	sq. m

		Labour (hr)	Plant compacting (hr)	Hardcore (tonne)	Cart away (cu. m)	Unit
D20.405	**Working space allowance to excavations, filling with excavated material, compacting with whacker in 250 mm layers**					
D20.410	Reduce levels, basements and the like	2.55	—	0.12	—	—	sq. m
D20.415	Pits	2.75	—	0.12	—	—	sq. m
D20.420	Trenches	2.75	—	0.12	—	—	sq. m
D20.425	Pile caps and ground beams between piles	3.67	—	0.12	—	—	sq. m
D20.450	**Working space allowance to excavations, filling with imported hardcore, compacting with whacker in 250 mm layers, disposing of surplus excavated material by removing from site**					
D20.455	Reduce levels, basements and the like	2.97	—	0.12	1.215	0.60	sq. m
D20.460	Pits	3.15	—	0.12	1.215	0.60	sq. m
D20.465	Trenches	3.15	—	0.12	1.215	0.60	sq. m
D20.470	Pile caps and ground beams between piles	4.22	—	0.12	1.215	0.60	sq. m

		Labour (hr)	Machine excavating (hr)		Wagon transport (hr)	Tipping charges (cu. m)	Unit
D20.505	**Disposal of excavated material (machine loading–JCB) from spoil heaps or sides of excavation**						
D20.510	Depositing on site in temporary spoil heaps, average distance from excavation						
	25 m	—	0.05	—	—	—	cu. m
	50 m	—	0.06	—	—	—	cu. m
	75 m	—	0.08	—	—	—	cu. m
	100 m	—	0.10	—	—	—	cu. m
D20.515	Depositing on site, spreading and levelling, average distance from excavation						
	25 m	—	0.07	—	—	—	cu. m
	50 m	—	0.08	—	—	—	cu. m
	75 m	—	0.11	—	—	—	cu. m
	100 m	—	0.13	—	—	—	cu. m
D20.520	Removing from site to tip and paying tipping charges, average distance from site						
	5 Km	—	0.07	—	0.17	1.00	cu. m
	10 Km	—	0.07	—	0.20	1.00	cu. m
	15 Km	—	0.07	—	0.25	1.00	cu. m
	20 Km	—	0.07	—	0.29	1.00	cu. m
D20.525	Loading, transporting and depositing in spoil heaps (multiple handling), average distance from excavations						
	25 m	—	0.05	—	—	—	cu. m
	50 m	—	0.06	—	—	—	cu. m
	75 m	—	0.08	—	—	—	cu. m
	100 m	—	0.10	—	—	—	cu. m

	Labour (hr)			Wagon transport (hr)	Tipping charges (cu. m)	Unit
D20.550 **Disposal of excavated material (hand loading) from spoil heaps or from side of excavations**						
D20.555 Depositing on site in temporary spoil heaps, average distance from excavation						
25 m	1.67	—	—	—	—	cu. m
50 m	1.75	—	—	—	—	cu. m
75 m	1.85	—	—	—	—	cu. m
100 m	1.96	—	—	—	—	cu. m
D20.560 Depositing, spreading and levelling on site, by hand, average distance from excavation						
25 m	2.50	—	—	—	—	cu. m
50 m	2.63	—	—	—	—	cu. m
75 m	2.77	—	—	—	—	cu. m
100 m	2.94	—	—	—	—	cu. m
D20.565 Removing from site to tip and paying tipping charges, average distance from site						
5 Km	1.50	—	—	0.57	1.00	cu. m
10 Km	1.50	—	—	0.61	1.00	cu. m
15 Km	1.50	—	—	0.63	1.00	cu. m
20 Km	1.50	—	—	0.67	1.00	cu. m
D20.570 Loading, transporting and depositing in spoil heaps (multiple handling), average distance from excavation						
25 m	1.67	—	—	—	—	cu. m
50 m	1.75	—	—	—	—	cu. m
75 m	1.85	—	—	—	—	cu. m
100 m	1.96	—	—	—	—	cu. m

E IN SITU CONCRETE

EA   Notes

1   The labour constants in this section are based on one labourer.

2   The labour constants in Section E10, columns 1–3 are for mixing and laying concrete, using the size of mixer indicated.

3   The labour constants in Section E10, column 4 are for the laying only of ready mixed concrete, delivered to site.

4   The cost of the mixer for site mixed concrete should be included in the cost of the concrete per cu. m (see Section EA) or as a separate plant cost in the preliminaries.

5   The average density of aggregates varies according to the particular material used and the moisture content. The bulk densities of sand, cement and aggregate used in Section EA – Concrete Mixes are as follows:

O.P. cement (in paper bags) – 1420 kg/m^3
Sand (fine aggregate)   – 1600 kg/m^3
Gravel (coarse aggregate)   – 1500 kg/m^3

CONCRETE MIXES

	Cement (tonne)	Sand (tonne)	Ballast/ Aggregate (tonne)	5/3.5 Mixer (hr)	7/5 Mixer (hr)	10/7 Mixer (hr)
Approximate weights of materials required to produce one cubic metre of fully compacted concrete including waste						
Nominal mixes – 20 mm aggregate						
1:3:6 nominal mix, 20 mm aggregate	0.213	0.72	1.35	1.33	0.89	0.67
1:2:4 nominal mix, 20 mm aggregate	0.304	0.69	1.29	1.33	0.89	0.67
1:1.5:3 nominal mix; 20 mm aggregate	0.387	0.65	1.23	1.33	0.89	0.67
1:1:2 nominal mix; 20 mm aggregate	0.533	0.60	1.13	1.33	0.89	0.67
Nominal mixes – All-in aggregate						
1:12 nominal mix; 20 mm down all-in ballast	0.164	—	2.22	1.33	0.89	0.67
1:9 nominal mix; 20 mm down all-in ballast	0.213	—	2.16	1.33	0.89	0.67
1:6 nominal mix; 20 mm down all-in ballast	0.304	—	2.06	1.33	0.89	0.67
Prescribed mixes CP.110						
25–75 mm slump; 20 mm aggregate, Grade 20–30, Zone 2, Fine aggregate						
Grade 7 (40% fine aggregate)	0.210	0.76	1.14	1.33	0.89	0.67
Grade 10 (40% fine aggregate)	0.240	0.74	1.11	1.33	0.89	0.67
Grade 15 (40% fine aggregate)	0.280	0.72	1.08	1.33	0.89	0.67
Grade 20 (35% fine aggregate)	0.320	0.63	1.08	1.33	0.89	0.67
Grade 25 (35% fine aggregate)	0.360	0.61	1.14	1.33	0.89	0.67
Grade 30 (35% fine aggregate)	0.400	0.60	1.10	1.33	0.89	0.67

E10   *IN SITU CONCRETE*

E10.005	**Plain in situ concrete**
E10.010	Foundations; filled into formwork
E10.015	Foundations; poured against earth faces
E10.020	Ground beams; filled into formwork
E10.025	Ground beams; poured against earth faces
E10.030	Isolated foundations; filled into formwork
E10.035	Isolated foundations; poured against earth faces
E10.040	Beds
E10.045	Filling to hollow walls
E10.050	Staircases
E10.055	Grouting in cement grout under stanchion bases
E10.105	**Reinforced in-situ concrete**
E10.110	Foundations; filled into formwork
E10.115	Foundations; poured against earth faces
E10.120	Ground beams; filled into formwork
E10.125	Ground beams; poured against earth faces
E10.130	Isolated foundations; filled into formwork
E10.135	Isolated foundations; poured against earth faces

E10.140	Beds
E10.145	Slabs
E10.150	Coffered and troughed slabs
E10.155	Walls
E10.160	Filling to hollow walls
E10.165	Beams
E10.170	Beam casings
E10.175	Columns
E10.180	Column casings
E10.185	Staircases
E10.190	Upstands

E10   IN SITU CONCRETE

	5/3.5 Mixer Labour (hr)	7/5 Mixer Labour (hr)	10/7 Mixer Labour (hr)	Ready mixed		Unit
				Labour (hr)	Concrete (cu. m)	
E10.005	**Plain in situ concrete**					
E10.010	Foundations, filled into formwork	5.33	4.01	3.34	1.33	1.00	cu. m
E10.015	Foundations, poured against earth faces						
	not exceeding 600 mm wide	5.76	4.33	3.61	1.44	1.08	cu. m
	600–900 mm wide	5.65	4.25	3.54	1.41	1.06	cu. m
	900–1200 mm wide	5.54	4.17	3.47	1.38	1.04	cu. m
	1200–1500 mm wide	5.49	4.13	3.44	1.37	1.03	cu. m
E10.020	Ground beams, filled into formwork	6.08	4.76	4.09	2.08	1.00	cu. m
E10.025	Ground beams, poured against earth faces						
	not exceeding 300 mm wide	6.81	5.33	4.58	2.33	1.12	cu. m
	300–600 mm wide	6.57	5.14	4.42	2.25	1.08	cu. m
	600–900 mm wide	6.38	5.00	4.29	2.18	1.05	cu. m
E10.030	Isolated foundations, filled into formwork	6.00	4.68	4.01	2.00	1.00	cu. m
E10.035	Isolated foundations, poured against earth faces						
	not exceeding 1.0 sq. m	6.60	5.15	4.41	2.20	1.10	cu. m
	1.00–2.25 sq. m	6.42	5.01	4.29	2.14	1.07	cu. m
	2.25–4.00 sq. m	6.30	4.91	4.21	2.10	1.05	cu. m
	4.00–9.00 sq. m	6.18	4.82	4.13	2.06	1.03	cu. m
E10.040	Beds						
	not exceeding 150 mm thick	5.43	4.11	3.44	1.43	1.00	cu. m
	150–450 mm thick	5.02	3.70	3.03	1.02	1.00	cu. m
	over 450 mm thick	4.71	3.39	2.72	0.71	1.00	cu. m
	Extra for beds laid to slopes						
	not exceeding 15 degrees from horizontal	0.25	0.25	0.25	0.25	—	cu. m
	over 15 degrees from horizontal	0.50	0.50	0.50	0.50	—	cu. m
E10.045	Filling to hollow walls						
	not exceeding 150 mm thick	7.33	6.01	5.34	3.33	1.00	cu. m
E10.050	Staircases	7.33	6.01	5.34	3.33	1.00	cu. m
E10.055	Grouting in cement grout under stanchion bases; 25 mm thick						
	not exceeding 600 × 600 mm	1.27	—	—	—	0.030	sq. m
	600 × 600 to 900 × 900 mm	0.88	—	—	—	0.022	sq. m
	900 × 900 to 1200 × 1200 mm	0.71	—	—	—	0.020	sq. m

		5/3.5 Mixer Labour (hr)	7/5 Mixer Labour (hr)	10/7 Mixer Labour (hr)	Ready mixed		
					Labour (hr)	Concrete (cu. m)	Unit
E10.105	**Reinforced in situ concrete**						
E10.110	Foundations, filled into formwork	5.67	4.35	3.68	1.67	1.00	cu. m
E10.115	Foundations, poured against earth faces						
	not exceeding 600 mm wide	6.12	4.70	3.97	1.80	1.08	cu. m
	600–900 mm wide	6.37	4.61	3.90	1.77	1.06	cu. m
	900–1200 mm wide	5.90	4.52	3.83	1.74	1.04	cu. m
	1200–1500 mm wide	5.84	4.48	3.79	1.72	1.03	cu. m
E10.120	Ground beams, filled into formwork	6.56	5.24	4.57	2.56	1.00	cu. m
E10.125	Ground beams, poured against earth faces						
	not exceeding 300 mm wide	7.35	5.87	5.12	2.87	1.12	cu. m
	300–600 mm wide	7.02	5.61	4.89	2.74	1.07	cu. m
	600–900 mm wide	6.89	5.50	4.80	2.69	1.05	cu. m
E10.130	Isolated foundations, filled into formwork	6.27	4.95	4.28	2.27	1.00	cu. m
E10.135	Isolated foundations, poured against earth faces						
	not exceeding 1.0 sq. m	6.90	5.45	4.71	2.50	1.10	cu. m
	1.00–2.25 sq. m	6.71	5.30	4.58	2.43	1.07	cu. m
	2.25–4.00 sq. m	6.58	5.20	4.49	2.38	1.05	cu. m
	4.00–9.00 sq. m	6.46	5.10	4.41	2.34	1.03	cu. m
E10.140	Beds						
	not exceeding 150 mm thick	5.85	4.53	3.86	1.85	1.00	cu. m
	150–450 mm thick	5.33	4.01	3.34	1.33	1.00	cu. m
	over 450 mm thick	4.93	3.61	2.94	0.93	1.00	cu. m
	Extra for beds laid to slopes						
	not exceeding 15 degrees from horizontal	0.25	0.25	0.25	0.25	—	cu. m
	over 15 degrees from horizontal	0.50	0.50	0.50	0.50	—	cu. m
E10.145	Slabs						
	not exceeding 150 mm thick	7.03	5.71	5.04	3.03	1.00	cu. m
	150–450 mm thick	6.13	4.81	4.14	2.13	1.00	cu. m
	over 450 mm thick	5.49	4.17	3.50	1.49	1.00	cu. m
E10.150	Coffered and troughed slabs						
	not exceeding 150 mm thick	7.03	5.71	5.04	3.03	1.00	cu. m
	150–450 mm thick	6.13	4.81	4.14	2.13	1.00	cu. m
	over 450 mm thick	5.49	4.17	3.50	1.49	1.00	cu. m
E10.155	Walls						
	not exceeding 150 mm thick	7.70	6.38	5.71	3.70	1.00	cu. m
	150–450 mm thick	7.23	5.91	5.24	3.23	1.00	cu. m
	over 450 mm thick	6.22	4.90	4.23	2.22	1.00	cu. m

		5/3.5 Mixer Labour (hr)	7/5 Mixer Labour (hr)	10/7 Labour (hr)	Ready mixed		Unit
					Labour (hr)	Concrete (cu. m)	
E10.160	Filling to hollow walls						
	not exceeding 150 mm thick	8.17	6.85	6.18	4.17	1.00	cu. m
E10.165	Beams						
	Isolated	8.00	6.68	6.01	4.00	1.00	cu. m
	Isolated deep	8.35	7.03	6.36	4.35	1.00	cu. m
	Attached deep	5.49	4.17	3.50	1.49	1.00	cu. m
E10.170	Beam casings						
	Isolated	8.27	6.90	6.21	4.13	1.00	cu. m
	Isolated deep	8.63	7.27	6.57	4.50	1.00	cu. m
	Attached deep	5.67	4.31	3.62	1.54	1.00	cu. m
E10.175	Columns	9.56	8.24	7.57	5.56	1.00	cu. m
E10.180	Column casings	9.88	8.56	7.89	5.88	1.00	cu. m
E10.185	Staircases	7.85	6.53	5.86	3.85	1.00	cu. m
E10.190	Upstands	8.35	7.03	6.36	4.35	1.00	cu. m

E30 *REINFORCEMENT FOR IN SITU CONCRETE*

E30 REINFORCEMENT FOR IN SITU CONCRETE

	Labour (hr)			Steel (tonne)	Tying wire (kg)	Unit

E30.005 **Bar reinforcement**

E30.010 Plain round steel bars, supplied cut, bent and labelled; straight and bent in any position

6 mm	90.00	—	—	1.025	0.17	tonne
8 mm	70.00	—	—	1.025	0.17	tonne
10 mm	60.00	—	—	1.025	0.14	tonne
12 mm	50.00	—	—	1.025	0.11	tonne
16 mm	40.00	—	—	1.025	0.11	tonne
20 mm	30.00	—	—	1.025	0.09	tonne
25 mm	30.00	—	—	1.025	0.09	tonne

E30.015 Links, stirrups, binders and special spacers in any position

6 mm	110.00	—	—	1.025	0.17	tonne
8 mm	90.00	—	—	1.025	0.17	tonne
10 mm	70.00	—	—	1.025	0.14	tonne
12 mm	60.00	—	—	1.025	0.11	tonne

E30.020 Hot rolled deformed high yield steel bars supplied cut, bent and labelled; straight and bent in any position

6 mm	90.00	—	—	1.025	0.17	tonne
8 mm	70.00	—	—	1.025	0.17	tonne
10 mm	60.00	—	—	1.025	0.14	tonne
12 mm	50.00	—	—	1.025	0.11	tonne
16 mm	40.00	—	—	1.025	0.11	tonne
20 mm	30.00	—	—	1.025	0.09	tonne
25 mm	30.00	—	—	1.025	0.09	tonne

E30.025 Links, stirrups, binders and special spacers in any position

6 mm	110.00	—	—	1.025	0.17	tonne
8 mm	90.00	—	—	1.025	0.17	tonne
10 mm	70.00	—	—	1.025	0.14	tonne
12 mm	60.00	—	—	1.025	0.11	tonne

	Labour (hr)			Mesh (sq. m)	Spacers (no.)	Unit

E30.105   **Steel fabric reinforcement**

E30.110   Fabric reinforcement with one width mesh side lap and one width mesh end lap; laid in ground slabs; reference

	Labour (hr)			Mesh (sq. m)	Spacers (no.)	Unit
A 98, 200 mm side, 200 mm end lap	0.04	—	—	1.22	6.88	sq. m
A 142, 200 mm side, 200 mm end lap	0.04	—	—	1.22	6.88	sq. m
A 193, 200 mm side, 200 mm end lap	0.05	—	—	1.22	6.88	sq. m
A 252, 200 mm side, 200 mm end lap	0.05	—	—	1.22	6.88	sq. m
A 393, 200 mm side, 200 mm end lap	0.07	—	—	1.22	6.88	sq. m
B 196, 100 mm side, 200 mm end lap	0.05	—	—	1.17	6.88	sq. m
B 283, 100 mm side, 200 mm end lap	0.05	—	—	1.17	6.88	sq. m
B 385, 100 mm side, 200 mm end lap	0.05	—	—	1.17	6.88	sq. m
B 503, 100 mm side, 200 mm end lap	0.07	—	—	1.17	6.88	sq. m
B 786, 100 mm side, 200 mm end lap	0.10	—	—	1.17	6.88	sq. m
B 1131, 100 mm side, 200 mm end lap	0.10	—	—	1.17	6.88	sq. m
C 283, 100 mm side, 400 mm end lap	0.05	—	—	1.24	6.88	sq. m
C 385, 100 mm side, 400 mm end lap	0.05	—	—	1.24	6.88	sq. m
C 503, 100 mm side, 400 mm end lap	0.05	—	—	1.24	6.88	sq. m
C 785, 100 mm side, 400 mm end lap	0.07	—	—	1.24	6.88	sq. m

E30.115   Fabric reinforcement with one width mesh side lap and one width mesh end lap; laid in suspended slabs, reference

	Labour (hr)			Mesh (sq. m)	Spacers (no.)	Unit
A 98, 200 mm side, 200 mm end lap	0.04	—	—	1.22	6.88	sq. m
A 142, 200 mm side, 200 mm end lap	0.04	—	—	1.22	6.88	sq. m
A 193, 200 mm side, 200 mm end lap	0.07	—	—	1.22	6.88	sq. m
A 252, 200 mm side, 200 mm end lap	0.07	—	—	1.22	6.88	sq. m
A 393, 200 mm side, 200 mm end lap	0.09	—	—	1.22	6.88	sq. m
B 196, 100 mm side, 200 mm end lap	0.06	—	—	1.17	6.88	sq. m
B 283, 100 mm side, 200 mm end lap	0.06	—	—	1.17	6.88	sq. m
B 385, 100 mm side, 200 mm end lap	0.06	—	—	1.17	6.88	sq. m
B 503, 100 mm side, 200 mm end lap	0.09	—	—	1.17	6.88	sq. m
B 785, 100 mm side, 200 mm end lap	0.13	—	—	1.17	6.88	sq. m
B 1131, 100 mm side, 200 mm end lap	0.13	—	—	1.17	6.88	sq. m
C 283, 100 mm side, 400 mm end lap	0.06	—	—	1.24	6.88	sq. m
C 385, 100 mm side, 400 mm end lap	0.06	—	—	1.24	6.88	sq. m
C 503, 100 mm side, 400 mm end lap	0.06	—	—	1.24	6.88	sq. m
C 785, 100 mm side, 400 mm end lap	0.09	—	—	1.24	6.88	sq. m

		Labour (hr)			Mesh (sq. m)	Spacers (no.)	Unit
E30.120	Fabric reinforcement with one width mesh side lap and one width mesh end lap; laid in walls; reference						
	A 98, 200 mm side, 200 mm end lap	0.14	—	—	1.22	6.88	sq. m
	A 142, 200 mm side, 200 mm end lap	0.14	—	—	1.22	6.88	sq. m
	A 193, 200 mm side, 200 mm end lap	0.18	—	—	1.22	6.88	sq. m
	A 252, 200 mm side, 200 mm end lap	0.18	—	—	1.22	6.88	sq. m
	A 393, 200 mm side, 200 mm end lap	0.21	—	—	1.22	6.88	sq. m
	B 196, 100 mm side, 200 mm end lap	0.18	—	—	1.17	6.88	sq. m
	B 283, 100 mm side, 200 mm end lap	0.18	—	—	1.17	6.88	sq. m
	B 385, 100 mm side, 200 mm end lap	0.18	—	—	1.17	6.88	sq. m
	B 503, 100 mm side, 200 mm end lap	0.21	—	—	1.17	6.88	sq. m
	B 785, 100 mm side, 200 mm end lap	0.31	—	—	1.17	6.88	sq. m
	B 1131, 100 mm side, 200 mm end lap	0.31	—	—	1.17	6.88	sq. m
	C 283, 100 mm side, 400 mm end lap	0.18	—	—	1.24	6.88	sq. m
	C 385, 100 mm side, 400 mm end lap	0.18	—	—	1.24	6.88	sq. m
	C 503, 100 mm side, 400 mm end lap	0.18	—	—	1.24	6.88	sq. m
	C 785, 100 mm side, 400 mm end lap	0.21	—	—	1.24	6.88	sq. m
E30.125	Wrapping fabric with one width mesh side and one width mesh end lap; in casings to steel columns and beams; reference						
	D 49, 100 mm side, 100 mm end lap	0.50	—	—	1.13	6.88	sq. m
	D 98, 100 mm side, 100 mm end lap	0.50	—	—	1.13	6.88	sq. m
E30.130	Add to above rates for one width mesh side and two width mesh end lap; reference						
	A 98, 200 mm side, 400 mm end lap	—	—	—	0.15	—	sq. m
	A 142, 200 mm side, 400 mm end lap	—	—	—	0.12	—	sq. m
	A 193, 200 mm side, 400 mm end lap	—	—	—	0.11	—	sq. m
	A 252, 200 mm side, 400 mm end lap	—	—	—	0.10	—	sq. m
	A 393, 200 mm side, 400 mm end lap	—	—	—	0.09	—	sq. m
	B 196, 100 mm side, 400 mm end lap	—	—	—	0.09	—	sq. m
	B 283, 100 mm side, 400 mm end lap	—	—	—	0.08	—	sq. m
	B 385, 100 mm side, 400 mm end lap	—	—	—	0.07	—	sq. m
	B 503, 100 mm side, 400 mm end lap	—	—	—	0.06	—	sq. m
	B 785, 100 mm side, 400 mm end lap	—	—	—	0.05	—	sq. m
	B 1131, 100 mm side, 400 mm end lap	—	—	—	0.05	—	sq. m
	C 283, 100 mm side, 800 mm end lap	—	—	—	0.09	—	sq. m
	C 385, 100 mm side, 800 mm end lap	—	—	—	0.07	—	sq. m
	C 503, 100 mm side, 800 mm end lap	—	—	—	0.06	—	sq. m
	C 785, 100 mm side, 800 mm end lap	—	—	—	0.05	—	sq. m

		Labour (hr)			Mesh (sq. m)	Spacers (no.)	Unit
E30.135	Add to above rates for two width mesh side and two width mesh end lap; reference						
	A 98, 400 mm side, 400 mm end lap	—	—	—	0.17	—	sq. m
	A 142, 400 mm side, 400 mm end lap	—	—	—	0.15	—	sq. m
	A 193, 400 mm side, 400 mm end lap	—	—	—	0.13	—	sq. m
	A 252, 400 mm side, 400 mm end lap	—	—	—	0.12	—	sq. m
	A 393, 400 mm side, 400 mm end lap	—	—	—	0.11	—	sq. m
	B 196, 200 mm side, 400 mm end lap	—	—	—	0.12	—	sq. m
	B 283, 200 mm side, 400 mm end lap	—	—	—	0.11	—	sq. m
	B 385, 200 mm side, 400 mm end lap	—	—	—	0.09	—	sq. m
	B 503, 200 mm side, 400 mm end lap	—	—	—	0.08	—	sq. m
	B 785, 200 mm side, 400 mm end lap	—	—	—	0.07	—	sq. m
	B 1131, 200 mm side, 400 mm end lap	—	—	—	0.07	—	sq. m
	C 283, 200 mm side, 800 mm end lap	—	—	—	0.14	—	sq. m
	C 385, 200 mm side, 800 mm end lap	—	—	—	0.12	—	sq. m
	C 503, 200 mm side, 800 mm end lap	—	—	—	0.11	—	sq. m
	C 785, 200 mm side, 800 mm end lap	—	—	—	0.11	—	sq. m
E30.140	Raking cutting fabric; reference						
	A 98	0.16	—	—	0.05	—	ln. m
	A 142	0.16	—	—	0.05	—	ln. m
	A 193	0.18	—	—	0.05	—	ln. m
	A 252	0.18	—	—	0.05	—	ln. m
	A 393	0.31	—	—	0.05	—	ln. m
	B 196	0.18	—	—	0.05	—	ln. m
	B 283	0.18	—	—	0.05	—	ln. m
	B 385	0.18	—	—	0.05	—	ln. m
	B 503	0.31	—	—	0.05	—	ln. m
	B 785	0.42	—	—	0.05	—	ln. m
	B 1131	0.42	—	—	0.05	—	ln. m
	C 283	0.18	—	—	0.05	—	ln. m
	C 385	0.18	—	—	0.05	—	ln. m
	C 503	0.18	—	—	0.05	—	ln. m
	C 785	0.31	—	—	0.05	—	ln. m
	D 49	0.16	—	—	0.05	—	ln. m
	D 98	0.16	—	—	0.05	—	ln. m

		Labour (hr)			Mesh (sq. m)	Spacers (no.)	Unit
E30.145	Circular cutting fabric; reference						
	A 98	0.21	—	—	0.05	—	ln. m
	A 142	0.21	—	—	0.05	—	ln. m
	A 193	0.31	—	—	0.05	—	ln. m
	A 252	0.31	—	—	0.05	—	ln. m
	A 395	0.42	—	—	0.05	—	ln. m
	B 196	0.31	—	—	0.05	—	ln. m
	B 283	0.31	—	—	0.05	—	ln. m
	B 385	0.31	—	—	0.05	—	ln. m
	B 503	0.42	—	—	0.05	—	ln. m
	B 785	0.63	—	—	0.05	—	ln. m
	B 1131	0.63	—	—	0.05	—	ln. m
	C 283	0.31	—	—	0.05	—	ln. m
	C 385	0.31	—	—	0.05	—	ln. m
	C 503	0.31	—	—	0.05	—	ln. m
	C 785	0.42	—	—	0.05	—	ln. m
	D 49	0.21	—	—	0.05	—	ln. m
	D 98	0.21	—	—	0.05	—	ln. m

E40 *DESIGNED JOINTS IN IN SITU CONCRETE*

E40.005	**Flexcel fibreboard joint filler**
E40.010	10 mm thick
E40.015	13 mm thick
E40.020	19 mm thick
E40.025	25 mm thick
E40.030	Extra; notching around reinforcing bars

E40   DESIGNED JOINTS IN IN SITU CONCRETE

		Labour (hr)				Flexcel (ln. m)	Unit
E40.005	**Flexcel fibreboard joint filler; fixing in position in joint; formwork**						
E40.010	10 mm thick						
	75 mm wide	0.02	—	—	—	1.05	ln. m
	100 mm wide	0.03	—	—	—	1.05	ln. m
	150 mm wide	0.04	—	—	—	1.05	ln. m
	200 mm wide	0.06	—	—	—	1.05	ln. m
	300 mm wide	0.08	—	—	—	1.05	ln. m
E40.015	13 mm thick; 75 mm wide	0.02	—	—	—	1.05	ln. m
	100 mm wide	0.03	—	—	—	1.05	ln. m
	150 mm wide	0.04	—	—	—	1.05	ln. m
	200 mm wide	0.06	—	—	—	1.05	ln. m
	300 mm wide	0.08	—	—	—	1.05	ln. m
E40.020	19 mm thick; 75 mm wide	0.02	—	—	—	1.05	ln. m
	100 mm wide	0.03	—	—	—	1.05	ln. m
	150 mm wide	0.04	—	—	—	1.05	ln. m
	200 mm wide	0.06	—	—	—	1.05	ln. m
	300 mm wide	0.09	—	—	—	1.05	ln. m
E40.025	25 mm thick;						
	75 mm wide	0.02	—	—	—	1.05	ln. m
	100 mm wide	0.03	—	—	—	1.05	ln. m
	150 mm wide	0.04	—	—	—	1.05	ln. m
	200 mm wide	0.06	—	—	—	1.05	ln. m
	300 mm wide	0.09	—	—	—	1.05	ln. m
E40.030	Extra; notching around reinforcing bars up to and including 25 mm diameter; any thickness						
	150 mm centres	0.38	—	—	—	—	ln. m
	200 mm centres	0.30	—	—	—	—	ln. m
	250 mm centres	0.24	—	—	—	—	ln. m
	300 mm centres	0.20	—	—	—	—	ln. m

E41   *WORKED FINISHES*

E41.005   **Worked finishes to surfaces of unset concrete**

E41.010   Tamping

E41.015   Power floating

E41.020   Trowelling

E41.025   Febco concrete hardening and dustproofing liquid

E41   WORKED FINISHES

	Labour (hr)	Plant (hr)		Febco liquid (litre)	Unit
E41.005	**Worked finishes to surfaces of unset concrete**				
E41.010	Tamping				
level	0.03	—	—	—	—	sq. m
to falls	0.04	—	—	—	—	sq. m
to crossfalls	0.05	—	—	—	—	sq. m
to cambers	0.07	—	—	—	—	sq. m
to slopes not exceeding 15 degree from horizontal	0.04	—	—	—	—	sq. m
E41.015	Power floating				
level	0.12	0.12	—	—	—	sq. m
to falls	0.15	0.15	—	—	—	sq. m
to crossfalls	0.20	0.20	—	—	—	sq. m
to cambers	0.22	0.22	—	—	—	sq. m
to slopes not exceeding 15 degree from horizontal	0.15	0.15	—	—	—	sq. m
E41.020	Trowelling				
level	0.17	—	—	—	—	sq. m
to falls	0.22	—	—	—	—	sq. m
to crossfalls	0.29	—	—	—	—	sq. m
to cambers	0.33	—	—	—	—	sq. m
to slopes not exceeding 15 degree from horizontal	0.22	—	—	—	—	sq. m
E41.025	Febco concrete hardening and dustproofing liquid; two coat work; brushing over floors or the like				
tamped finish concrete surfaces					
not exceeding 150 mm wide	0.03	—	—	—	0.043	ln. m
150–300 mm wide	0.05	—	—	—	0.086	ln. m
over 300 mm wide	0.16	—	—	—	0.286	sq. m
trowelled finish concrete surfaces;					
not exceeding 150 mm wide	0.02	—	—	—	0.030	ln. m
150–300 mm wide	0.04	—	—	—	0.060	ln. m
over 300 mm wide	0.13	—	—	—	0.200	sq. m

F MASONRY

FA   Notes

I   The labour constants in this section are based on two craftsmen and one labourer, two and one
        bricklaying gang.

2   Add the cost per hour of two craftsmen bricklayers and one labourer, divide the total by three and
        multiply the 'gang rate' by the constants given.

3   The cost of a mortar mixer should be included in the cost of mortar per cu. m (see Section FA–Mortar
        Mixes), or as a separate plant cost in the preliminaries.

4   The constants for special facing bricks in Section F10 are 'full value' constants. Where described as 'extra
        over' in the bill of quantities, the rate for the brickwork on which they occur should be deducted.

MORTAR MIXES

Approximate weights of materials required to
produce one cubic metre of mortar, including
waste

	Cement (tonne)	Walcrete (tonne)	Lime (tonne)	Sand (tonne)	Admix (litre)	5/3.5 Mixer (hr)
Cement, Lime and Sand Mortar						
1:1:6 cement, lime, sand	0.237	—	0.12	1.60	—	1.33
1:2:9 cement, lime, sand	0.158	—	0.16	1.60	—	1.33
1:3:9 cement, lime, sand	0.146	—	0.22	1.48	—	1.33
Cement and Sand Mortar						
1:3 cement, sand	0.473	—	—	1.60	2.37	1.33
1:4 cement, sand	0.379	—	—	1.71	1.90	1.33
1:6 cement, sand	0.270	—	—	1.83	1.35	1.33
Cement and Sand Grout						
1:1 cement, sand	0.947	—	—	1.07	—	1.33
1:2 cement, sand	0.631	—	—	1.42	—	1.33
Walcrete Mortar						
1:4 walcrete, sand	—	0.379	—	1.71	—	1.33
1:5 walcrete, sand	—	0.316	—	1.83	—	1.33

	Cement (tonne)	L.S.M. (tonne)				5/3.5 Mixer (hr)
Lime:Sand Mortars						
1:0.25:3 cement, LSM (1:3)	0.473	1.93	—	—	—	1.33
1:1:6 cement, LSM (1:6)	0.270	2.20	—	—	—	1.33
1:2:9 cement, LSM (1:9)	0.189	2.31	—	—	—	1.33

F10   *BRICK/BLOCK WALLING*

F10.005	**Common bricks, 215 × 102.5 × 65 mm**
F10.010	Walls
F10.015	Walls; built curved; 2.5–5.0 m radius on face
F10.020	Walls; built curved; over 5.0 m radius on face
F10.025	Isolated piers, casings or chimney stacks
F10.030	Projections
F10.035	Arches; flat
F10.040	Arches; segmental
F10.045	Arches; semicircular
F10.050	Closing cavities with common brickwork 102.5 mm thick

F10.105	**Common bricks, 215 × 102.5 × 65 mm, bucket handle or weather struck pointing as work proceeds**
F10.110	Walls; facing and pointing one side
F10.115	Walls; facing and pointing both sides
F10.120	Walls; built curved; facing and pointing one side; 2.5–5.0 m radius on face
F10.125	Walls; built curved; facing and pointing one side over 5.0 m radius on face
F10.130	Walls; built curved; facing and pointing both sides; 2.5–5.0 m radius on face
F10.135	Walls; built curved; facing and pointing both sides; over 5.0 m radius on face
F10.140	Isolated piers, casings or chimney stacks; facing and pointing both sides
F10.145	Projections
F10.150	Arches; flat
F10.155	Arches; segmental
F10.160	Arches; semicircular
F10.165	Plain bands; stretcher bond; projecting from face of wall; facing and pointing margins
F10.170	Plain bands; stretcher bond; sunk; set back from face of wall; facing and pointing margins
F10.175	Brick-on-end bands; flush with face of wall
F10.180	Brick-on-end bands; projecting from face of wall; facing and pointing margins
F10.185	Brick-on-end bands; sunk; set back from face of wall; facing and pointing margins

F10.205	**Common bricks, 215 × 102.5 × 65 mm, flush pointing as work proceeds**
F10.210	Walls; facing and pointing one side
F10.215	Walls; facing and pointing both sides
F10.220	Walls; built curved; facing and pointing one side; 2.5–5.0 m radius on face
F10.225	Walls; built curved; facing and pointing one side; over 5.0 m radius on face
F10.230	Walls; built curved; facing and pointing both sides; 2.5–5.0 m radius on face
F10.235	Walls; built curved; facing and pointing both sides; over 5.0 m radius on face
F10.240	Isolated piers, casings or chimney stacks; facing and pointing both sides
F10.245	Projections
F10.250	Arches; flat
F10.255	Arches; segmental
F10.260	Arches; semicircular
F10.265	Plain bands; stretcher bond; projecting from face of wall; facing and pointing margins
F10.270	Plain bands; stretcher bond; sunk; set back from face of wall; facing and pointing margins
F10.275	Brick-on-end bands; flush with face of wall
F10.280	Brick-on-end bands; projecting from face of wall; facing and pointing margins
F10.285	Brick-on-end bands; sunk; set back from face of wall; facing and pointing margins

F10.305	**Facing bricks as specified, 215 × 102.5 × 65 mm, in mortar as specified; bucket handle or weather struck pointing as work proceeds**
F10.310	Walls; facing and pointing one side
F10.315	Walls; facing and pointing both sides
F10.320	Walls; built curved; facing and pointing one side; 2.5–5.0 m radius on face
F10.325	Walls; built curved; facing and pointing one side; over 5.0 m radius on face
F10.330	Walls; built curved; facing and pointing both sides; 2.5–5.0 m radius on face
F10.335	Walls; built curved; facing and pointing both sides; over 5.0 m radius on face
F10.340	Isolated piers, casings or chimney stacks; facing and pointing both sides
F10.345	Projections
F10.350	Arches; flat
F10.355	Arches; segmental
F10.360	Arches; semicircular
F10.365	Closing cavities with facing brickwork, 102.5 mm thick
F10.370	Plain bands; stretcher bond; projecting from face of wall; facing and pointing margins
F10.375	Plain bands; stretcher bond; sunk; set back from face of wall; facing and pointing margins
F10.380	Brick-on-end bands; flush with face of wall

F10.385   Brick-on-end bands; projecting from face of wall; facing and pointing margins

F10.390   Brick-on-end bands; sunk; set back from face of wall; facing and pointing margins

F10.395   Corbels; pointing face, each side and returns

F10.396   Copings; 215 × 112.5 mm, brick-on-edge; facing and pointing top and two sides

F10.397   Sills; 225 × 112.5 mm, brick-on-edge; projecting 50 mm from wall; set weathering; facing and pointing returns

F10.405	**Special bricks to BS.4729:1971 in facings**
F10.410	Half round coping; type 2.2.1
F10.415	Saddleback coping; type 2.2.2
F10.420	Single bullnose; type 2.3.1.1 or 2.3.1.2
F10.425	Single bullnose; type 2.3.3.1 or 2.3.3.2
F10.430	Single bullnose; type 2.3.8.1 or 2.3.8.2
F10.435	Single bullnose; type 2.3.10.1 or 2.3.10.2
F10.440	Double bullnose; type 2.3.2.1 or 2.3.2.2
F10.445	Double bullnose; type 2.3.4.1 or 2.3.4.2
F10.450	Bullnose header; type 2.3.5.1 or 2.3.5.2
F10.455	Bullnose stretcher; type 2.3.6.1 or 2.3.6.2
F10.460	Bullnose internal return; type 2.3.9.1 or 2.3.9.2
F10.465	Bullnose external return; type 2.3.11.1 or 2.3.11.2
F10.470	Bullnose double header; type 2.3.13.1 or 2.3.13.2
F10.475	Bullnose double stretcher; type 2.3.14.1 or 2.3.14.2
F10.480	Bullnose on end (cownose); type 2.3.16
F10.485	Squint on edge; type 2.4.1.1, 2.4.1.2 or 2.4.1.3
F10.490	External angle; type 2.4.2.1, 2.4.2.2 or 2.4.2.3
F10.495	Internal angle; type 2.4.3.1, 2.4.3.2 or 2.4.3.3
F10.505	Birdsmouth; type 2.4.4.1, 2.4.4.2 or 2.4.4.3
F10.510	Single cant; type 2.4.5.1 or 2.4.5.2

F10.515   Double cant; type 2.4.6.1 or 2.4.6.2

F10.520   Plinth header; type 2.5.2.1 or 2.5.2.2

F10.525   Plinth stretcher; type 2.5.3.1 or 2.5.3.2

F10.530   Plinth internal return (long); type 2.5.4.1 or 2.5.4.2

F10.535   Plinth internal return (short); type 2.5.5.1 or 2.5.5.2

F10.540   Plinth internal angle; type 2.5.6.1 or 2.5.6.2

F10.545   Plinth external return; type 2.5.7.1 or 2.5.7.2

F10.550   Plinth external angle; type 2.5.8.1 or 2.5.8.2

F10.605	**Type A precast concrete blocks, natural aggregate, face size 440 × 215 mm**
F10.610	Walls
F10.615	Walls; built curved; 2.5–5.0 m radius on face
F10.620	Walls; built curved; over 5.0 m radius on face
F10.625	Isolated piers, casings or chimney stacks
F10.630	Projections
F10.635	Closing cavities with blockwork
F10.640	Extra for bonding ends of blockwork to new brickwork; extra material for bonding
F10.645	**Type A precast concrete blocks, natural aggregate, face size 440 × 215 mm; fair face pointing as work proceeds**
F10.650	Walls; facing and pointing one side
F10.655	Walls; facing and pointing both sides
F10.660	Walls; built curved; facing and pointing one side; 2.5–5.0 m radius on face
F10.665	Walls; built curved; facing and pointing one side; over 5.0 m radius on face
F10.670	Walls; built curved; facing and pointing both sides; 2.5–5.0 m radius on face
F10.675	Walls; built curved; facing and pointing both sides; over 5.0 m radius on face
F10.680	Isolated piers, casings or chimney stacks; facing and pointing one side
F10.685	Isolated piers, casings or chimney stacks; facing and pointing both sides
F10.690	Projections; facing and pointing one side

F10.705	**Type B precast concrete blocks, furnace clinker aggregate, face size 440 × 215 mm**
F10.710	Walls
F10.715	Walls; built curved; 2.5–5.0 m radius on face
F10.720	Walls; built curved; over 5.0 m radius on face
F10.725	Isolated piers, casings or chimney stacks
F10.730	Projections
F10.735	Closing cavities with blockwork
F10.740	Extra for bonding ends of blockwork to new brickwork; extra material for bonding

F10.745	**Type B precast concrete blocks, furnace clinker aggregate, face size 440 × 215 mm; fair face pointing as work proceeds.**
F10.750	Walls; facing and pointing one side
F10.755	Walls; facing and pointing both sides
F10.760	Walls; built curved; facing and pointing one side; 2.5–5.0 m radius on face
F10.765	Walls; built curved; facing and pointing one side; over 5.0 m radius on face
F10.770	Walls; built curved; facing and pointing both sides; 2.5–5.0 m radius on face
F10.775	Walls; built curved; facing and pointing both sides; over 5.0 m radius on face
F10.780	Isolated piers, casings or chimney stacks; facing and pointing one side
F10.785	Isolated piers, casings or chimney stacks; facing and pointing both sides
F10.790	Projections; facing and pointing one side

F10.805	**Precast concrete blocks, autoclaved, aerated; face size 440×215 mm**
F10.810	Walls
F10.815	Walls; built curved; 2.5–5.0 m radius on face
F10.820	Walls; built curved; over 5.0 m radius on face
F10.825	Isolated piers, casings or chimney stacks
F10.830	Projections
F10.835	Closing cavities with blockwork
F10.840	Extra for bonding ends of blockwork to new brickwork; extra material for bonding
F10.845	**Precast concrete blocks, autoclaved, aerated; face size 440×215 mm; fair face pointing as work proceeds**
F10.850	Walls; facing and pointing one side
F10.855	Walls; facing and pointing both sides
F10.860	Walls; built curved; facing and pointing one side; 2.5–5.0 m radius on face
F10.865	Walls; built curved; facing and pointing one side; over 5.0 m radius on face
F10.870	Walls; built curved; facing and pointing both sides; 2.5–5.0 m radius on face
F10.875	Walls; built curved; facing and pointing both sides; over 5.0 m radius on face
F10.880	Isolated piers, casings or chimney stacks; facing and pointing one side
F10.885	Isolated piers, casings or chimney stacks; facing and pointing both sides
F10.890	Projections; facing and pointing one side

F10   BRICK / BLOCK WALLING

	Labour (hr)		Bricks (no.)	Mortar No frog (cu. m)	One frog (cu. m)	Unit
F10.005	**Common bricks, 215 × 103 × 65 mm, in any mortar**					
F10.010	Walls					
103 mm, stretcher bond	1.48	—	60.44	0.018	0.032	sq. m
215 mm, English bond	2.54	—	120.87	0.045	0.074	sq. m
328 mm, English bond	2.96	—	181.31	0.073	0.117	sq. m
440 mm, English bond	3.56	—	241.74	0.101	0.159	sq. m
F10.015	Walls; built curved; 2.5–5.0 m radius on face					
103 mm, stretcher bond	2.47	—	60.44	0.018	0.032	sq. m
215 mm, English bond	4.23	—	120.87	0.045	0.074	sq. m
328 mm, English bond	4.94	—	181.31	0.073	0.117	sq. m
440 mm, English bond	5.93	—	241.74	0.101	0.159	sq. m
F10.020	Walls; built curved; over 5.0 m radius on face					
103 mm, Stretcher bond	2.22	—	60.44	0.018	0.032	sq. m
215 mm, English bond	3.78	—	120.87	0.045	0.074	sq. m
328 mm, English bond	4.44	—	181.31	0.073	0.117	sq. m
440 mm, English bond	5.31	—	241.74	0.101	0.159	sq. m
F10.025	Isolated piers, casings or chimney stacks					
103 mm; stretcher bond	1.85	—	60.44	0.018	0.032	sq. m
215 mm; English bond	3.17	—	120.87	0.045	0.074	sq. m
328 mm; English bond	3.70	—	181.31	0.073	0.117	sq. m
440 mm; English bond	4.44	—	241.74	0.101	0.159	sq. m
553 mm; English bond	5.09	—	302.18	0.119	0.191	sq. m
665 mm; English bond	5.55	—	362.61	0.146	0.233	sq. m
F10.030	Projections					
215 wide × 113 mm thick	0.46	—	13.00	0.006	0.009	ln. m
328 wide × 113 mm thick	0.71	—	19.80	0.008	0.013	ln. m
440 wide × 113 mm thick	0.95	—	26.60	0.010	0.016	ln. m
215 wide × 225 mm thick	0.80	—	26.00	0.012	0.018	ln. m
328 wide × 225 mm thick	1.21	—	39.60	0.017	0.026	ln. m
440 wide × 225 mm thick	1.67	—	54.40	0.022	0.035	ln. m
440 wide × 328 mm thick	1.94	—	81.60	0.034	0.054	ln. m

		Labour (hr)	Ties (no.)	Bricks (no.)	Mortar		Unit
					No frog (cu. m)	One frog (cu. m)	
F10.035	Arches; flat						
	225 mm wide on face, brick-on-end						
	113 mm on soffit	0.52	—	13.67	0.004	0.007	ln. m
	113 mm wide on face, brick-on-edge						
	113 mm on soffit	0.47	—	13.67	0.004	0.007	ln. m
	225 mm on soffit	0.52	—	13.67	0.004	0.007	ln. m
F10.040	Arches; segmental						
	225 mm wide on face, brick-on-edge; in two half brick rings						
	113 mm on soffit	0.65	—	27.33	0.008	0.014	ln. m
	225 mm on soffit	0.75	—	27.33	0.008	0.014	ln. m
F10.045	Arches; semicircular						
	225 mm wide on face, brick-on-edge; in two half brick rings						
	113 mm on soffit	1.03	—	27.33	0.008	0.014	ln. m
	225 mm on soffit	1.15	—	27.33	0.008	0.014	ln. m
F10.050	Closing cavities with common brickwork 103 thick						
	at jambs or ends of walls	0.35	4.89	4.533	0.001	0.002	ln. m
	at sills	0.46	—	4.533	0.001	0.002	ln. m
	at tops of walls	0.23	—	4.533	0.001	0.002	ln. m

		Labour (hr)		Bricks (no.)	Mortar		Unit
					No frog (cu. m)	One frog (cu. m)	
F10.105	**Common bricks, 215 × 103 × 65 mm, in any mortar; bucket handle or weather struck pointing as work proceeds**						
F10.110	Walls; facing and pointing one side						
	103 mm, stretcher bond	1.86	—	60.44	0.018	0.032	sq. m
	215 mm, English bond	2.95	—	120.87	0.045	0.074	sq. m
F10.115	Walls; facing and pointing both sides						
	103 mm, stretcher bond	2.24	—	60.44	0.018	0.032	sq. m
	215 mm, English bond	3.36	—	120.87	0.045	0.074	sq. m
F10.120	Walls; built curved, facing and pointing one side; 2.5–5.0 m radius on face						
	103 mm, stretcher bond	2.85	—	60.44	0.018	0.032	sq. m
	215 mm, English bond	4.64	—	120.87	0.045	0.074	sq. m
	328 mm, English bond	5.35	—	181.31	0.073	0.117	sq. m
	440 mm, English bond	6.34	—	241.74	0.101	0.159	sq. m
F10.125	Walls; built curved, facing and pointing one side; over 5.0 m radius on face						
	103 mm, stretcher bond	2.60	—	60.44	0.018	0.032	sq. m
	215 mm, English bond	4.19	—	120.87	0.045	0.074	sq. m
	328 mm, English bond	4.85	—	181.31	0.073	0.117	sq. m
	440 mm, English bond	5.72	—	241.74	0.101	0.159	sq. m
F10.130	Walls; built curved, facing and pointing both sides; 2.5–5.0 m radius on face						
	103 mm, stretcher bond	3.23	—	60.44	0.018	0.032	sq. m
	215 mm, English bond	5.05	—	120.87	0.045	0.074	sq. m
	328 mm, English bond	5.76	—	181.31	0.073	0.117	sq. m
	440 mm, English bond	6.75	—	241.74	0.101	0.159	sq. m
F10.135	Walls; built curved, facing and pointing both sides; over 5.0 m radius on face						
	103 mm, stretcher bond	2.98	—	60.44	0.018	0.032	sq. m
	215 mm, English bond	4.60	—	120.87	0.045	0.074	sq. m
	328 mm, English bond	5.26	—	181.31	0.073	0.117	sq. m
	440 mm, English bond	6.13	—	241.74	0.101	0.159	sq. m
F10.140	Isolated piers, casings or chimney stacks, facing and pointing both sides						
	103 mm, stretcher bond	2.61	—	60.44	0.018	0.032	sq. m
	215 mm, English bond	3.99	—	120.87	0.045	0.074	sq. m

		Labour (hr)		Bricks (no.)	Mortar		Unit
					No frog (cu. m)	One frog (cu. m)	
F10.145	Projections						
	215 wide × 113 mm thick	0.63	—	13.00	0.006	0.009	ln. m
	328 wide × 113 mm thick	0.92	—	19.80	0.008	0.013	ln. m
	440 wide × 113 mm thick	1.20	—	26.60	0.010	0.016	ln. m
	215 wide × 225 mm thick	1.05	—	26.00	0.012	0.018	ln. m
	328 wide × 225 mm thick	1.51	—	39.60	0.017	0.026	ln. m
	440 wide × 225 mm thick	2.01	—	54.40	0.022	0.035	ln. m
	440 wide × 328 mm thick	2.36	—	81.60	0.034	0.054	ln. m
F10.150	Arches; flat						
	225 mm wide on face, brick-on-end;						
	113 mm on soffit	0.65	—	13.67	0.004	0.007	ln. m
	113 mm wide on face, brick-on-edge						
	113 mm on soffit	0.56	—	13.67	0.004	0.007	ln. m
	225 mm on soffit	0.65	—	13.67	0.004	0.007	ln. m
F10.155	Arches; segmental						
	225 mm wide on face, brick-on-edge; in						
	two half brick rings						
	113 mm on soffit	0.78	—	27.33	0.008	0.014	ln. m
	225 mm on soffit	0.92	—	27.33	0.008	0.014	ln. m
F10.160	Arches; semicircular						
	225 mm wide on face, brick-on-edge; in						
	two half brick rings						
	113 mm on soffit	1.16	—	27.33	0.008	0.014	ln. m
	225 mm on soffit	1.32	—	27.33	0.008	0.014	ln. m
F10.165	Plain bands; stretcher bond; projecting from face of wall; facing and pointing margins						
	horizontal						
	65 mm wide	0.20	—	4.53	0.001	0.002	ln. m
	140 mm wide	0.41	—	9.06	0.003	0.004	ln. m
	215 mm wide	0.62	—	13.60	0.004	0.007	ln. m
	raking						
	65 mm wide	0.21	—	4.53	0.001	0.002	ln. m
	140 mm wide	0.44	—	9.06	0.003	0.004	ln. m
	215 mm wide	0.67	—	13.60	0.004	0.007	ln. m

		Labour (hr)		Bricks (no.)	Mortar		Unit
					No frog (cu. m)	One frog (cu. m)	
F10.170	Plain bands; stretcher bond; sunk; set back from face of wall; facing and pointing margins						
	horizontal						
	85 mm wide	0.22	—	4.53	0.002	0.003	ln. m
	160 mm wide	0.42	—	9.06	0.003	0.005	ln. m
	235 mm wide	0.62	—	13.60	0.004	0.008	ln. m
	raking						
	85 mm wide	0.24	—	4.53	0.002	0.003	ln. m
	160 mm wide	0.45	—	9.06	0.003	0.005	ln. m
	235 mm wide	0.66	—	13.60	0.004	0.008	ln. m
F10.175	Brick-on-end bands; flush with face of wall; 225 mm wide						
	horizontal	0.71	—	13.60	0.004	0.008	ln. m
	raking	0.96	—	13.60	0.004	0.008	ln. m
F10.180	Brick-on-end bands; projecting from face of wall; facing and pointing margins; 215 mm wide						
	horizontal	2.03	—	13.60	0.004	0.008	ln. m
	raking	2.63	—	13.60	0.004	0.008	ln. m
F10.185	Brick-on-end bands; sunk; set back from face of wall; facing and pointing margins; 65 mm wide						
	horizontal	0.94	—	4.53	0.001	0.002	ln. m
	raking	1.12	—	4.53	0.001	0.002	ln. m

		Labour (hr)		Bricks (no.)	Mortar		Unit
					No frog (cu. m)	One frog (cu. m)	
F10.205	**Common bricks, 215 × 103 × 65 mm, in any mortar; flush pointing as work proceeds**						
F10.210	Walls; facing and pointing one side						
	103 mm, stretcher bond	1.83	—	60.44	0.018	0.032	sq. m
	215 mm, English bond	2.91	—	120.87	0.045	0.074	sq. m
F10.215	Walls; facing and pointing both sides						
	103 mm, stretcher bond	2.18	—	60.44	0.018	0.032	sq. m
	215 mm, English bond	3.28	—	120.87	0.045	0.074	sq. m
F10.220	Walls; built curved, facing and pointing one side; 2.5–5.0 m radius on face						
	103 mm, stretcher bond	2.82	—	60.44	0.018	0.032	sq. m
	215 mm, English bond	4.60	—	120.87	0.045	0.074	sq. m
	328 mm, English bond	5.31	—	181.31	0.073	0.117	sq. m
	440 mm, English bond	6.30	—	241.74	0.101	0.159	sq. m
F10.225	Walls; built curved, facing and pointing one side; over 5.0 m radius on face						
	103 mm, stretcher bond	2.57	—	60.44	0.018	0.032	sq. m
	215 mm, English bond	4.15	—	120.87	0.045	0.074	sq. m
	328 mm, English bond	4.81	—	181.31	0.073	0.117	sq. m
	440 mm, English bond	5.68	—	241.74	0.101	0.159	sq. m
F10.230	Walls; built curved, facing and pointing both sides; 2.5–5.0 m radius on face						
	103 mm, stretcher bond	3.17	—	60.44	0.018	0.032	sq. m
	215 mm, English bond	4.97	—	120.87	0.045	0.074	sq. m
	328 mm, English bond	5.68	—	181.31	0.073	0.117	sq. m
	440 mm, English bond	6.67	—	241.74	0.101	0.159	sq. m
F10.235	Walls; built curved, facing and pointing both sides; over 5.0 m radius on face						
	103 mm, stretcher bond	2.92	—	60.44	0.018	0.032	sq. m
	215 mm, English bond	4.52	—	120.87	0.045	0.074	sq. m
	328 mm, English bond	5.18	—	181.31	0.073	0.117	sq. m
	440 mm, English bond	6.05	—	241.74	0.101	0.159	sq. m
F10.240	Isolated piers, casings or chimney stacks; facing and pointing both sides						
	103 mm, stretcher bond	2.55	—	60.44	0.018	0.032	sq. m
	215 mm, English bond	3.91	—	120.87	0.045	0.074	sq. m

		Labour (hr)		Bricks (no.)	Mortar No frog (cu. m)	One frog (cu. m)	Unit
F10.245	Projections						
	215 wide × 113 mm thick	0.61	—	13.00	0.006	0.009	ln. m
	328 wide × 113 mm thick	0.90	—	19.80	0.008	0.013	ln. m
	440 wide × 113 mm thick	1.18	—	26.60	0.010	0.016	ln. m
	215 wide × 225 mm thick	1.03	—	26.00	0.012	0.018	ln. m
	328 wide × 225 mm thick	1.48	—	39.60	0.017	0.026	ln. m
	440 wide × 225 mm thick	1.98	—	54.40	0.022	0.035	ln. m
	440 wide × 328 mm thick	2.32	—	81.60	0.034	0.054	ln. m
F10.250	Arches; flat						
	225 mm wide on face, brick-on-end;						
	113 mm on soffit	0.64	—	13.67	0.004	0.007	ln. m
	113 mm wide on face, brick-on-edge						
	113 mm on soffit	0.55	—	13.67	0.004	0.007	ln. m
	225 mm on soffit	0.64	—	13.67	0.004	0.007	ln. m
F10.255	Arches; segmental						
	225 mm wide on face, brick-on-edge; in						
	two half brick rings						
	113 mm on soffit	0.77	—	27.33	0.008	0.014	ln. m
	225 mm on soffit	0.91	—	27.33	0.008	0.014	ln. m
F10.260	Arches; semicircular						
	225 mm wide on face, brick-on-edge; in						
	two half brick rings						
	113 mm on soffit	1.15	—	27.33	0.008	0.014	ln. m
	225 mm on soffit	1.31	—	27.33	0.008	0.014	ln. m
F10.265	Plain bands; stretcher bond; projecting from face of wall; facing and pointing margins						
	horizontal;						
	65 mm wide	0.21	—	4.53	0.001	0.002	ln. m
	140 mm wide	0.41	—	9.06	0.003	0.004	ln. m
	215 mm wide	0.61	—	13.60	0.004	0.007	ln. m
	raking						
	65 mm wide	0.21	—	4.53	0.001	0.002	ln. m
	140 mm wide	0.44	—	9.06	0.003	0.004	ln. m
	215 mm wide	0.66	—	13.60	0.004	0.007	ln. m

		Labour (hr)		Bricks (no.)	Mortar		Unit
					No frog (cu. m)	One frog (cu. m)	
F10.270	Plain bands; stretcher bond; sunk; set back from face of wall; facing and pointing margins						
	horizontal						
	85 mm wide	0.22	—	4.53	0.002	0.003	ln. m
	160 mm wide	0.41	—	9.06	0.003	0.005	ln. m
	235 mm wide	0.61	—	13.60	0.004	0.008	ln. m
	raking						
	85 mm wide	0.24	—	4.53	0.002	0.003	ln. m
	160 mm wide	0.44	—	9.06	0.003	0.005	ln. m
	235 mm wide	0.65	—	13.60	0.004	0.008	ln. m
F10.275	Brick-on-end bands; flush with face of wall; 225 mm wide						
	horizontal	0.70	—	13.60	0.004	0.007	ln. m
	raking	0.95	—	13.60	0.004	0.007	ln. m
F10.280	Brick-on-end bands; projecting from face of wall; facing and pointing margins; 215 mm wide						
	horizontal	2.02	—	13.60	0.004	0.007	ln. m
	raking	2.62	—	13.60	0.004	0.007	ln. m
F10.285	Brick-on-end bands; sunk; set back from face of wall; facing and pointing margins; 65 mm wide						
	horizontal	0.94	—	4.53	0.001	0.002	ln. m
	raking	1.12	—	4.53	0.001	0.002	ln. m

		Labour (hr)		Bricks (no.)	Mortar		Unit
					No frog (cu. m)	One frog (cu. m)	
F10.305	**Facing bricks as specified, 215 × 103 × 65 mm, in mortar as specified; bucket handle or weather struck pointing as work proceeds**						
F10.310	Walls; facing and pointing one side						
	103 mm, stretcher bond	2.12	—	60.44	0.018	0.032	sq. m
	215 mm, English bond	3.37	—	120.87	0.045	0.074	sq. m
F10.315	Walls; facing and pointing both sides						
	103 mm, stretcher bond	2.50	—	60.44	0.018	0.032	sq. m
	215 mm, English bond	3.78	—	120.87	0.045	0.074	sq. m
F10.320	Walls; built curved, facing and pointing one side; 2.5–5.0 m radius on face						
	103 mm, stretcher bond	3.34	—	60.44	0.018	0.032	sq. m
	215 mm, English bond	5.35	—	120.87	0.045	0.074	sq. m
F10.325	Walls; built curved, facing and pointing one side; over 5.0 m radius on face						
	103 mm, stretcher bond	2.99	—	60.44	0.018	0.032	sq. m
	215 mm, English bond	4.85	—	120.87	0.045	0.074	sq. m
F10.330	Walls; built curved, facing and pointing both sides; 2.5–5.0 m radius on face;						
	103 mm, stretcher bond	3.72	—	60.44	0.018	0.032	sq. m
	215 mm, English bond	5.76	—	120.87	0.045	0.074	sq. m
F10.335	Walls; built curved, facing and pointing both sides; over 5.0 m radius on face						
	103 mm, stretcher bond	3.37	—	60.44	0.018	0.032	sq. m
	215 mm, English bond	5.26	—	120.87	0.045	0.074	sq. m
F10.340	Isolated piers, casings or chimney stacks; facing and pointing both sides;						
	103 mm, stretcher bond	3.13	—	60.44	0.018	0.032	sq. m
	215 mm, English bond	4.73	—	120.87	0.045	0.074	sq. m
F10.345	Projections;						
	215 wide × 113 mm thick	0.66	—	13.00	0.006	0.009	ln. m
	328 wide × 113 mm thick	1.01	—	19.80	0.008	0.013	ln. m
	440 wide × 113 mm thick	1.35	—	26.60	0.010	0.016	ln. m

		Labour (hr)	Ties (no.)	Bricks (no.)	Mortar		Unit
					No frog (cu. m)	One frog (cu. m)	
F10.350	Flat arches						
	225 mm wide on face, brick-on-end						
	113 mm on soffit	0.86	—	13.67	0.004	0.007	ln. m
	113 mm wide on face, brick-on-edge;						
	113 mm on soffit	0.78	—	13.67	0.004	0.007	ln. m
	225 mm on soffit	0.86	—	13.67	0.004	0.007	ln. m
F10.355	Segmental arches						
	225 mm wide on face, brick-on-edge; in						
	two half brick rings						
	113 mm on soffit	1.09	—	27.33	0.008	0.014	ln. m
	225 mm on soffit	1.25	—	27.33	0.008	0.014	ln. m
F10.360	Semicircular arches						
	225 mm wide on face, brick-on-edge; in						
	two half brick rings						
	113 mm on soffit	1.72	—	27.33	0.008	0.014	ln. m
	225 mm on soffit	1.92	—	27.33	0.008	0.014	ln. m
F10.365	Closing cavities with facing brickwork; 103 mm thick						
	at jambs or ends of walls	0.67	4.89	4.53	0.001	0.002	ln. m
	at sills	0.46	—	4.53	0.001	0.002	ln. m
	at tops of walls	0.33	—	4.53	0.001	0.002	ln. m
F10.370	Plain bands; stretcher bond; projecting from face of wall; facing and pointing margins						
	horizontal						
	65 mm wide	0.22	—	4.56	0.001	0.002	ln. m
	140 mm wide	0.43	—	9.11	0.003	0.005	ln. m
	215 mm wide	0.65	—	13.67	0.004	0.007	ln. m
	raking						
	65 mm wide	0.23	—	4.56	0.001	0.002	ln. m
	140 mm wide	0.46	—	9.11	0.003	0.005	ln. m
	215 mm wide	0.69	—	13.67	0.004	0.007	ln. m
F10.375	Plain bands; stretcher bond; sunk; set back from face of wall; facing and pointing margins						
	horizontal						
	85 mm wide	0.20	—	4.56	0.001	0.002	ln. m
	160 mm wide	0.41	—	9.11	0.003	0.005	ln. m
	235 mm wide	0.13	—	13.67	0.004	0.007	ln. m
	raking						
	85 mm wide	0.22	—	4.56	0.001	0.002	ln. m
	160 mm wide	0.43	—	9.11	0.003	0.005	ln. m
	235 mm wide	0.65	—	13.67	0.004	0.007	ln. m

		Labour (hr)		Bricks (no.)	Mortar		Unit
					No frog (cu. m)	One frog (cu. m)	
F10.380	Brick-on-end bands; flush with face of wall; 225 mm wide						
	horizontal	0.68	—	13.67	0.004	0.007	ln. m
	raking	0.86	—	13.67	0.004	0.007	ln. m
F10.385	Brick-on-end bands; projecting from face of wall; facing and pointing margins; 215 mm wide						
	horizontal	1.65	—	13.67	0.004	0.007	ln. m
	raking	2.08	—	13.67	0.004	0.007	ln. m
F10.390	Brick-on-end bands; sunk; set back from face of wall; facing and pointing margins; 65 mm wide						
	horizontal	1.06	—	13.67	0.004	0.007	ln. m
	raking	1.19	—	13.67	0.004	0.007	ln. m
F10.395	Corbels; pointing face, each side and returns; 215 × 165 mm total projection × 300 mm deep; four courses, three courses projecting 55 mm beyond course below	1.25	—	7.18	0.003	0.004	no.
F10.396	Copings; 215 × 113 mm, brick-on-edge; facing and pointing top and two sides	0.98	—	13.60	0.005	0.008	ln. m
F10.397	Sills; 225 × 113 mm, brick-on-edge; projecting 50 mm from wall; set weathering; facing and pointing returns	0.86	—	13.60	0.005	0.008	ln. m

		Labour (hr)		Bricks (no.)	Mortar		Unit
					No frog (cu. m)	One frog (cu. m)	
F10.405	**Special bricks to BS.4729:1971 in facings**						
F10.410	Half round coping; 305 × 153 × 65 mm on edge, type 2.2.1						
	straight	0.75	—	13.67	0.003	0.003	ln. m
	curved; radius 2.5–5.0 m	1.20	—	13.67	0.003	0.003	ln. m
	curved; radius over 5.0 m	1.13	—	13.67	0.003	0.003	ln. m
F10.415	Saddleback coping; 305 × 153 × 65 mm on edge, type 2.2.2						
	straight	0.75	—	13.67	0.003	0.003	ln. m
	curved; radius 2.5–5.0 m	1.20	—	13.67	0.003	0.003	ln. m
	curved; radius over 5.0 m	1.13	—	13.67	0.003	0.003	ln. m
F10.420	Single bullnose; 215 × 103 × 65 mm on edge; type 2.3.1.1 or 2.3.1.2						
	straight	0.38	—	13.67	0.002	0.002	ln. m
	curved; radius 2.5–5.0 m	0.61	—	13.67	0.002	0.002	ln. m
	curved; radius over 5.0 m	0.57	—	13.67	0.002	0.002	ln. m
F10.425	Single bullnose, 215 × 103 × 65 mm stop end; on edge; type 2.3.3.1 or 2.3.3.2	0.06	—	1.03	0.001	0.001	no.
F10.430	Single bullnose, 215 × 103 × 65 mm internal return on edge; type 2.3.8.1 or 2.3.8.2	0.06	—	1.03	0.001	0.001	no.
F10.435	Single bullnose, 215 × 103 × 65 mm external return on edge; type 2.3.10.1 or 2.3.10.2	0.06	—	1.03	0.001	0.001	no.
F10.440	Double bullnose; 215 × 103 × 65 mm on edge; type 2.3.2.1 or 2.3.2.2						
	straight	0.38	—	13.67	0.002	0.002	ln. m
	curved; radius 2.5–5.0 m	0.61	—	13.67	0.002	0.002	ln. m
	curved; radius over 5.0 m	0.57	—	13.67	0.002	0.002	ln. m
F10.445	Double bullnose, stop end; on edge; 215 × 103 × 65 mm; type 2.3.4.1 or 2.3.4.2	0.06	—	1.03	0.001	0.001	no.
F10.450	Bullnose header; 215 × 103 × 65 mm on flat; type 2.3.5.1 or 2.3.5.2						
	straight	0.25	—	9.07	0.003	0.003	ln. m
	curved; radius 2.5–5.0 m	0.40	—	9.07	0.003	0.003	ln. m
	curved; radius over 5.0 m	0.38	—	9.07	0.003	0.003	ln. m

		Labour (hr)		Bricks (no.)	Mortar		Unit
					No frog (cu. m)	One frog (cu. m)	
F10.455	Bullnose stretcher; 215 × 103 × 65 mm on flat; type 2.3.6.1 or 2.3.6.2						
	straight	0.13	—	4.56	0.002	0.002	ln. m
	curved; radius 2.5–5.0 m	0.21	—	4.56	0.002	0.002	ln. m
	curved; radius over 5.0 m	0.20	—	4.56	0.002	0.002	ln. m
F10.460	Bullnose internal return on flat; 215 × 103 × 65 mm; type 2.3.9.1 or 2.3.9.2	0.06	—	1.03	0.001	0.001	no.
F10.465	Bullnose external return on flat; 215 × 103 × 65 mm; type 2.3.11.1 or 2.3.11.2	0.06	—	1.03	0.001	0.001	no.
F10.470	Bullnose double header; 215 × 103 × 65 mm on flat; type 2.3.13.1 or 2.3.13.2						
	straight	0.25	—	9.07	0.003	0.003	ln. m
	curved; radius 2.5–5.0 m	0.40	—	9.07	0.003	0.003	ln. m
	curved; radius over 5.0 m	0.38	—	9.07	0.003	0.003	ln. m
F10.475	Bullnose double stretcher on flat; 215 × 103 × 65 mm; type 2.3.14.1 or 2.3.14.2						
	straight	0.13	—	4.56	0.002	0.002	ln. m
	curved; radius 2.5–5.0 m	0.21	—	4.56	0.002	0.002	ln. m
	curved; radius over 5.0 m	0.20	—	4.56	0.002	0.002	ln. m
F10.480	Bullnose on end (cownose); on end; 215 × 103 × 65 mm; type 2.3.16						
	straight	0.38	—	13.67	0.002	0.002	ln. m
	curved; radius 2.5–5.0 m	0.61	—	13.67	0.002	0.002	ln. m
	curved; radius over 5.0 m	0.57	—	13.67	0.002	0.002	ln. m
F10.485	Squint on edge; 164 × 103 × 65 mm; type 2.4.1.1, 2.4.1.2 or 2.4.1.3						
	straight	0.38	—	13.67	0.002	0.002	ln. m
	curved; radius 2.5–5.0 m	0.61	—	13.67	0.002	0.002	ln. m
	curved; radius over 5.0 m	0.57	—	13.67	0.002	0.002	ln. m
F10.490	External angle; 164 × 103 × 65 mm; on flat; type 2.4.2.1, 2.4.2.2 or 2.4.2.3	0.06	—	1.03	0.001	0.001	no.
F10.495	Internal angle; 164 × 103 × 65 mm; on flat; type 2.4.3.1, 2.4.3.2 or 2.4.3.3	0.06	—	1.03	0.001	0.001	no.

	Labour (hr)		Bricks (no.)	Mortar		Unit
				No frog (cu. m)	One frog (cu. m)	
F10.505	Birdsmouth, 215 × 103 × 65 mm; on flat; type 2.4.4.1, 2.4.4.2 or 2.4.4.3	0.06	—	1.03	0.001	0.001	no.
F10.510	Single cant, 215 × 103 × 65 mm on edge; type 2.4.5.1 or 2.4.5.2						
	straight	0.38	—	13.67	0.002	0.002	ln. m
	curved; radius 2.5–5.0 m	0.61	—	13.67	0.002	0.002	ln. m
	curved; radius over 5.0 m	0.57	—	13.67	0.002	0.002	ln. m
F10.515	Double cant, 215 × 103 × 65 mm on edge; type 2.4.6.1 or 2.4.6.2						
	straight	0.38	—	13.67	0.002	0.002	ln. m
	curved; radius 2.5–5.0 m	0.61	—	13.67	0.002	0.002	ln. m
	curved; radius over 5.0 m	0.57	—	13.67	0.002	0.002	ln. m
F10.520	Plinth header; 215 × 103 × 65 mm on flat; type 2.5.2.1 or 2.5.2.2						
	straight	0.25	—	9.07	0.003	0.003	ln. m
	curved; radius 2.5–5.0 m	0.40	—	9.07	0.003	0.003	ln. m
	curved; radius over 5.0 m	0.38	—	9.07	0.003	0.003	ln. m
F10.525	Plinth stretcher, 215 × 103 × 65 mm on flat; type 2.5.3.1 or 2.5.3.2						
	straight	0.13	—	4.56	0.002	0.002	ln. m
	curved; radius 2.5–5.0 m	0.21	—	4.56	0.002	0.002	ln. m
	curved; radius over 5.0 m	0.20	—	4.56	0.002	0.002	ln. m
F10.530	Plinth internal return (long); 215 × 103 × 65 mm on flat; type 2.5.4.1 or 2.5.4.2	0.06	—	1.03	0.001	0.001	no.
F10.535	Plinth internal return (short); 215 × 103 × 65 mm on flat; type 2.5.5.1 or 2.5.5.2	0.06	—	1.03	0.001	0.001	no.
F10.540	Plinth internal angle; 215 × 103 × 65 mm on flat; type 2.5.6.1 or 2.5.6.2	0.06	—	1.03	0.001	0.001	no.
F10.545	Plinth external return; 215 × 103 × 65 mm on flat; type 2.5.7.1 or 2.5.7.2	0.06	—	1.03	0.001	0.001	no.
F10.550	Plinth external angle; 215 × 103 × 65 mm on flat; type 2.5.8.1 or 2.5.8.2	0.06	—	1.03	0.001	0.001	no.

	Labour (hr)			Blocks (sq. m)	Mortar (cu. m)	Unit
F10.605	**Type A precast concrete blocks, natural aggregate, face size 440 × 215 mm**					
F10.610	Walls					
	solid blocks					
	100 mm thick	1.00	—	—	1.050	0.007	sq. m
	140 mm thick	1.20	—	—	1.050	0.009	sq. m
	190 mm thick	1.36	—	—	1.050	0.013	sq. m
	hollow blocks					
	100 mm thick	0.94	—	—	1.075	0.007	sq. m
	140 mm thick	1.15	—	—	1.075	0.009	sq. m
	190 mm thick	1.30	—	—	1.075	0.013	sq. m
	215 mm thick	1.43	—	—	1.075	0.014	sq. m
F10.615	Walls; built curved; 2.5–5 m radius on face					
	solid blocks					
	100 mm thick	1.67	—	—	1.050	0.007	sq. m
	140 mm thick	2.00	—	—	1.050	0.009	sq. m
	190 mm thick	2.27	—	—	1.050	0.013	sq. m
	hollow blocks					
	100 mm thick	1.57	—	—	1.075	0.007	sq. m
	140 mm thick	1.92	—	—	1.075	0.009	sq. m
	190 mm thick	2.17	—	—	1.075	0.013	sq. m
	215 mm thick	2.39	—	—	1.075	0.014	sq. m
F10.620	Walls; built curved; over 5.0 m radius on face					
	solid blocks					
	100 mm thick	1.50	—	—	1.050	0.007	sq. m
	140 mm thick	1.80	—	—	1.050	0.009	sq. m
	190 mm thick	2.04	—	—	1.050	0.013	sq. m
	hollow blocks					
	100 mm thick	1.41	—	—	1.075	0.007	sq. m
	140 mm thick	1.73	—	—	1.075	0.009	sq. m
	190 mm thick	1.95	—	—	1.075	0.013	sq. m
	215 mm thick	2.15	—	—	1.075	0.014	sq. m
F10.625	Isolated piers, casings or chimney stacks					
	solid blocks					
	100 mm thick	1.18	—	—	1.050	0.007	sq. m
	140 mm thick	1.42	—	—	1.050	0.009	sq. m
	190 mm thick	1.60	—	—	1.050	0.013	sq. m
	hollow blocks					
	100 mm thick	1.10	—	—	1.075	0.007	sq. m
	140 mm thick	1.36	—	—	1.075	0.009	sq. m
	190 mm thick	1.53	—	—	1.075	0.013	sq. m
	215 mm thick	1.69	—	—	1.075	0.014	sq. m

		Labour (hr)			Blocks (sq. m)	Mortar (cu. m)	Unit
F10.630	Projections						
	solid blocks						
	215 wide × 100 mm thick	0.31	—	—	0.226	0.002	ln. m
	440 wide × 100 mm thick	0.64	—	—	0.462	0.003	ln. m
	215 wide × 140 mm thick	0.37	—	—	0.226	0.002	ln. m
	440 wide × 140 mm thick	0.77	—	—	0.462	0.004	ln. m
	215 wide × 190 mm thick	0.42	—	—	0.226	0.003	ln. m
	440 wide × 190 mm thick	0.87	—	—	0.462	0.006	ln. m
	hollow blocks						
	215 wide × 100 mm thick	0.29	—	—	0.226	0.002	ln. m
	440 wide × 100 mm thick	0.60	—	—	0.462	0.003	ln. m
	215 wide × 140 mm thick	0.36	—	—	0.226	0.002	ln. m
	440 wide × 140 mm thick	0.73	—	—	0.462	0.004	ln. m
	215 wide × 190 mm thick	0.41	—	—	0.226	0.003	ln. m
	440 wide × 190 mm thick	0.83	—	—	0.462	0.006	ln. m
	215 wide × 215 mm thick	0.45	—	—	0.226	0.003	ln. m
	440 wide × 215 mm thick	0.91	—	—	0.462	0.006	ln. m
F10.635	Closing cavities with blockwork; 100 mm thick						
	at jambs or ends of walls	0.44	—	—	0.079	0.001	ln. m
	at sills	0.31	—	—	0.079	0.001	ln. m
	at tops of walls	0.23	—	—	0.079	0.001	ln. m
F10.640	Extra for bonding ends of blockwork to new brickwork; extra material for bonding						
	solid blocks						
	100 mm thick	—	—	—	0.053	0.001	ln. m
	140 mm thick	—	—	—	0.053	0.001	ln. m
	190 mm thick	—	—	—	0.053	0.001	ln. m
	hollow blocks						
	100 mm thick	—	—	—	0.054	0.001	ln. m
	140 mm thick	—	—	—	0.054	0.001	ln. m
	190 mm thick	—	—	—	0.054	0.001	ln. m
	215 mm thick	—	—	—	0.054	0.001	ln. m

		Labour (hr)			Blocks (sq. m)	Mortar (cu. m)	Unit
F10.645	**Type A precast concrete blocks, natural aggregate, face size 440×215 mm; fair face pointing as work proceeds**						
F10.650	Walls; facing and pointing one side						
	solid blocks						
	100 mm thick	1.14	—	—	1.050	0.007	sq. m
	140 mm thick	1.34	—	—	1.050	0.009	sq. m
	190 mm thick	1.50	—	—	1.050	0.013	sq. m
	hollow blocks						
	100 mm thick	1.08	—	—	1.075	0.007	sq. m
	140 mm thick	1.29	—	—	1.075	0.009	sq. m
	190 mm thick	1.44	—	—	1.075	0.013	sq. m
	215 mm thick	1.57	—	—	1.075	0.014	sq. m
F10.655	Walls; facing and pointing both sides						
	solid blocks						
	100 mm thick	1.28	—	—	1.050	0.007	sq. m
	140 mm thick	1.48	—	—	1.050	0.009	sq. m
	190 mm thick	1.64	—	—	1.050	0.013	sq. m
	hollow blocks						
	100 mm thick	1.22	—	—	1.075	0.007	sq. m
	140 mm thick	1.43	—	—	1.075	0.009	sq. m
	190 mm thick	1.58	—	—	1.075	0.013	sq. m
	215 mm thick	1.71	—	—	1.075	0.014	sq. m
F10.660	Walls; built curved; facing and pointing one side; 2.5–5.0 m radius on face						
	solid blocks						
	100 mm thick	1.95	—	—	1.050	0.007	sq. m
	140 mm thick	2.28	—	—	1.050	0.009	sq. m
	190 mm thick	2.55	—	—	1.050	0.013	sq. m
	hollow blocks						
	100 mm thick	1.85	—	—	1.075	0.007	sq. m
	140 mm thick	2.20	—	—	1.075	0.009	sq. m
	190 mm thick	2.45	—	—	1.075	0.013	sq. m
	215 mm thick	2.67	—	—	1.075	0.014	sq. m

		Labour (hr)			Blocks (sq. m)	Mortar (cu. m)	Unit
F10.665	Walls; built curved; facing and pointing one side; over 5.0 m radius on face						
	solid blocks						
	100 mm thick	1.78	—	—	1.050	0.007	sq. m
	140 mm thick	2.08	—	—	1.050	0.009	sq. m
	190 mm thick	2.32	—	—	1.050	0.013	sq. m
	hollow blocks						
	100 mm thick	1.69	—	—	1.075	0.007	sq. m
	140 mm thick	2.01	—	—	1.075	0.009	sq. m
	190 mm thick	2.23	—	—	1.075	0.013	sq. m
	215 mm thick	2.43	—	—	1.075	0.014	sq. m
F10.670	Walls; built curved; facing and pointing both sides; 2.5–5.0 m radius on face						
	solid blocks						
	100 mm thick	2.09	—	—	1.050	0.007	sq. m
	140 mm thick	2.42	—	—	1.050	0.009	sq. m
	190 mm thick	2.69	—	—	1.050	0.013	sq. m
	hollow blocks						
	100 mm thick	1.99	—	—	1.075	0.007	sq. m
	140 mm thick	2.34	—	—	1.075	0.009	sq. m
	190 mm thick	2.59	—	—	1.075	0.013	sq. m
	215 mm thick	2.81	—	—	1.075	0.014	sq. m
F10.675	Walls; built curved; facing and pointing both sides; over 5.0 m radius on face						
	solid blocks						
	100 mm thick	1.92	—	—	1.050	0.007	sq. m
	140 mm thick	2.22	—	—	1.050	0.009	sq. m
	190 mm thick	2.46	—	—	1.050	0.013	sq. m
	hollow blocks						
	100 mm thick	1.83	—	—	1.075	0.007	sq. m
	140 mm thick	2.15	—	—	1.075	0.009	sq. m
	190 mm thick	2.37	—	—	1.075	0.013	sq. m
	215 mm thick	2.57	—	—	1.075	0.014	sq. m

		Labour (hr)			Blocks (sq. m)	Mortar (cu. m)	Unit
F10.680	Isolated piers, casings or chimney stacks; facing and pointing one side						
	solid blocks						
	100 mm thick	1.46	—	—	1.050	0.007	sq. m
	140 mm thick	1.70	—	—	1.050	0.009	sq. m
	190 mm thick	1.88	—	—	1.050	0.013	sq. m
	hollow blocks;						
	100 mm thick	1.38	—	—	1.075	0.007	sq. m
	140 mm thick	1.64	—	—	1.075	0.009	sq. m
	190 mm thick	1.81	—	—	1.075	0.013	sq. m
	215 mm thick	1.97	—	—	1.075	0.014	sq. m
F10.685	Isolated piers, casings or chimney stacks; facing and pointing both sides						
	solid blocks						
	100 mm thick	1.60	—	—	1.050	0.007	sq. m
	140 mm thick	1.84	—	—	1.050	0.009	sq. m
	190 mm thick	2.02	—	—	1.050	0.013	sq. m
	hollow blocks;						
	100 mm thick	1.52	—	—	1.075	0.007	sq. m
	140 mm thick	1.78	—	—	1.075	0.009	sq. m
	190 mm thick	1.95	—	—	1.075	0.013	sq. m
	215 mm thick	2.11	—	—	1.075	0.014	sq. m
F10.690	Projections; facing and pointing one side						
	solid blocks						
	215 wide × 100 mm thick	0.34	—	—	0.226	0.002	ln. m
	440 wide × 100 mm thick	0.70	—	—	0.462	0.003	ln. m
	215 wide × 140 mm thick	0.40	—	—	0.226	0.002	ln. m
	440 wide × 140 mm thick	0.83	—	—	0.462	0.004	ln. m
	215 wide × 190 mm thick	0.45	—	—	0.226	0.003	ln. m
	440 wide × 190 mm thick	0.93	—	—	0.462	0.006	ln. m
	hollow blocks						
	215 wide × 100 mm thick	0.32	—	—	0.226	0.002	ln. m
	440 wide × 100 mm thick	0.66	—	—	0.462	0.003	ln. m
	215 wide × 140 mm thick	0.39	—	—	0.226	0.002	ln. m
	440 wide × 140 mm thick	0.79	—	—	0.462	0.004	ln. m
	215 wide × 190 mm thick	0.44	—	—	0.226	0.003	ln. m
	440 wide × 190 mm thick	0.89	—	—	0.462	0.006	ln. m
	215 wide × 215 mm thick	0.48	—	—	0.226	0.003	ln. m
	440 wide × 215 mm thick	0.97	—	—	0.462	0.006	ln. m

	Labour (hr)			Blocks (sq. m)	Mortar (cu. m)	Unit
F10.705	**Type B precast concrete blocks, furnace clinker aggregate, face size 440 × 215.0 mm**					
F10.710	Walls					
	solid blocks					
	75 mm thick	0.75	—	—	1.075	0.005	sq. m
	100 mm thick	0.79	—	—	1.050	0.007	sq. m
	140 mm thick	0.96	—	—	1.050	0.009	sq. m
	190 mm thick	1.09	—	—	1.050	0.013	sq. m
	215 mm thick	1.22	—	—	1.050	0.014	sq. m
	hollow blocks					
	100 mm thick	0.75	—	—	1.075	0.007	sq. m
	140 mm thick	0.92	—	—	1.075	0.009	sq. m
	190 mm thick	1.04	—	—	1.075	0.013	sq. m
	215 mm thick	1.15	—	—	1.075	0.014	sq. m
F10.715	Walls; built curved; 2.5–5.0 m radius on face					
	solid blocks					
	75 mm thick	1.25	—	—	1.075	0.005	sq. m
	100 mm thick	1.32	—	—	1.050	0.007	sq. m
	140 mm thick	1.60	—	—	1.050	0.009	sq. m
	190 mm thick	1.82	—	—	1.050	0.013	sq. m
	215 mm thick	2.04	—	—	1.050	0.014	sq. m
	hollow blocks					
	100 mm thick	1.25	—	—	1.075	0.007	sq. m
	140 mm thick	1.54	—	—	1.075	0.009	sq. m
	190 mm thick	1.74	—	—	1.075	0.013	sq. m
	215 mm thick	1.92	—	—	1.075	0.014	sq. m
F10.720	Walls; built curved; over 5.0 m radius on face					
	solid blocks					
	75 mm thick	1.13	—	—	1.075	0.005	sq. m
	100 mm thick	1.19	—	—	1.050	0.007	sq. m
	140 mm thick	1.44	—	—	1.050	0.009	sq. m
	190 mm thick	1.64	—	—	1.050	0.013	sq. m
	215 mm thick	1.83	—	—	1.050	0.014	sq. m
	hollow blocks					
	100 mm thick	1.13	—	—	1.075	0.007	sq. m
	140 mm thick	1.38	—	—	1.075	0.009	sq. m
	190 mm thick	1.56	—	—	1.075	0.013	sq. m
	215 mm thick	1.73	—	—	1.075	0.014	sq. m

		Labour (hr)			Blocks (sq. m)	Mortar (cu. m)	Unit
F10.725	Isolated piers, casings or chimney stacks						
	solid blocks						
	75 mm thick	0.88	—	—	1.050	0.005	sq. m
	100 mm thick	0.93	—	—	1.050	0.007	sq. m
	140 mm thick	1.13	—	—	1.050	0.009	sq. m
	190 mm thick	1.28	—	—	1.050	0.013	sq. m
	215 mm thick	1.76	—	—	1.050	0.014	sq. m
	hollow blocks						
	100 mm thick	0.88	—	—	1.075	0.007	sq. m
	140 mm thick	1.09	—	—	1.075	0.009	sq. m
	190 mm thick	1.23	—	—	1.075	0.013	sq. m
	215 mm thick	1.35	—	—	1.075	0.014	sq. m
F10.730	Projections						
	solid blocks						
	215 wide × 100 mm thick	0.25	—	—	0.226	0.002	ln. m
	440 wide × 100 mm thick	0.50	—	—	0.462	0.003	ln. m
	215 wide × 140 mm thick	0.30	—	—	0.226	0.002	ln. m
	440 wide × 140 mm thick	0.61	—	—	0.462	0.004	ln. m
	215 wide × 190 mm thick	0.34	—	—	0.226	0.003	ln. m
	440 wide × 190 mm thick	0.70	—	—	0.462	0.006	ln. m
	hollow blocks						
	215 wide × 100 mm thick	0.23	—	—	0.226	0.002	ln. m
	440 wide × 100 mm thick	0.48	—	—	0.462	0.003	ln. m
	215 wide × 140 mm thick	0.29	—	—	0.226	0.002	ln. m
	440 wide × 140 mm thick	0.59	—	—	0.462	0.004	ln. m
	215 wide × 190 mm thick	0.32	—	—	0.226	0.003	ln. m
	440 wide × 190 mm thick	0.66	—	—	0.462	0.006	ln. m
	215 wide × 215 mm thick	0.36	—	—	0.226	0.003	ln. m
	440 wide × 215 mm thick	0.73	—	—	0.462	0.006	ln. m
F10.735	Closing cavities with blockwork; 100 mm thick						
	at jambs or ends of walls	0.33	—	—	0.079	0.001	ln. m
	at sills	0.23	—	—	0.079	0.001	ln. m
	at tops of walls	0.17	—	—	0.079	0.001	ln. m

		Labour (hr)			Blocks (sq. m)	Mortar (cu. m)	Unit
F10.740	Extra for bonding ends of blockwork to new brickwork; extra material for bonding						
	solid blocks						
	75 mm thick	—	—	—	0.054	0.001	ln. m
	100 mm thick	—	—	—	0.053	0.001	ln. m
	140 mm thick	—	—	—	0.053	0.001	ln. m
	190 mm thick	—	—	—	0.053	0.001	ln. m
	215 mm thick	—	—	—	0.053	0.001	ln. m
	hollow blocks						
	100 mm thick	—	—	—	0.054	0.001	ln. m
	140 mm thick	—	—	—	0.054	0.001	ln. m
	190 mm thick	—	—	—	0.054	0.001	ln. m
	215 mm thick	—	—	—	0.054	0.001	ln. m

	Labour (hr)			Blocks (sq. m)	Mortar (cu. m)	Unit
F10.745	**Type B Precast Concrete blocks, furnace clinker aggregate, face size 440×215 mm; fair face pointing as work proceeds**					
F10.750	Walls; facing and pointing one side					
	solid blocks					
	75 mm thick	0.89	—	—	1.075	0.005	sq. m
	100 mm thick	0.93	—	—	1.050	0.007	sq. m
	140 mm thick	1.10	—	—	1.050	0.009	sq. m
	190 mm thick	1.23	—	—	1.050	0.013	sq. m
	215 mm thick	1.36	—	—	1.050	0.014	sq. m
	hollow blocks					
	100 mm thick	0.89	—	—	1.075	0.007	sq. m
	140 mm thick	1.06	—	—	1.075	0.009	sq. m
	190 mm thick	1.18	—	—	1.075	0.013	sq. m
	215 mm thick	1.29	—	—	1.075	0.014	sq. m
F10.755	Walls; facing and pointing both sides					
	solid blocks					
	75 mm thick	1.03	—	—	1.075	0.005	sq. m
	100 mm thick	1.07	—	—	1.050	0.007	sq. m
	140 mm thick	1.24	—	—	1.050	0.009	sq. m
	190 mm thick	1.37	—	—	1.050	0.013	sq. m
	215 mm thick	1.50	—	—	1.050	0.014	sq. m
	hollow blocks					
	100 mm thick	1.03	—	—	1.075	0.007	sq. m
	140 mm thick	1.20	—	—	1.075	0.009	sq. m
	190 mm thick	1.32	—	—	1.075	0.013	sq. m
	215 mm thick	1.43	—	—	1.075	0.014	sq. m
F10.760	Walls; built curved; facing and pointing one side; 2.5–5.0 m radius on face					
	solid blocks					
	75 mm thick	1.39	—	—	1.075	0.005	sq. m
	100 mm thick	1.46	—	—	1.050	0.007	sq. m
	140 mm thick	1.74	—	—	1.050	0.009	sq. m
	190 mm thick	1.96	—	—	1.050	0.013	sq. m
	215 mm thick	2.18	—	—	1.050	0.014	sq. m
	hollow blocks					
	100 mm thick	1.39	—	—	1.075	0.007	sq. m
	140 mm thick	1.68	—	—	1.075	0.009	sq. m
	190 mm thick	1.88	—	—	1.075	0.013	sq. m
	215 mm thick	2.06	—	—	1.075	0.014	sq. m

		Labour (hr)			Blocks (sq. m)	Mortar (cu. m)	Unit
F10.765	Walls; built curved; facing and pointing one side; over 5.00 m radius on face						
	solid blocks						
	75 mm thick	1.27	—	—	1.075	0.005	sq. m
	100 mm thick	1.33	—	—	1.050	0.007	sq. m
	140 mm thick	1.58	—	—	1.050	0.009	sq. m
	190 mm thick	1.78	—	—	1.050	0.013	sq. m
	215 mm thick	1.97	—	—	1.050	0.014	sq. m
	hollow blocks						
	100 mm thick	1.27	—	—	1.075	0.007	sq. m
	140 mm thick	1.52	—	—	1.075	0.009	sq. m
	190 mm thick	1.70	—	—	1.075	0.013	sq. m
	215 mm thick	1.87	—	—	1.075	0.014	sq. m
F10.770	Walls; built curved; facing and pointing both sides; 2.5–5.0 m radius on face						
	solid blocks						
	75 mm thick	1.53	—	—	1.075	0.005	sq. m
	100 mm thick	1.60	—	—	1.050	0.007	sq. m
	140 mm thick	1.88	—	—	1.050	0.009	sq. m
	190 mm thick	2.10	—	—	1.050	0.013	sq. m
	215 mm thick	2.32	—	—	1.050	0.014	sq. m
	hollow blocks						
	100 mm thick	1.53	—	—	1.075	0.007	sq. m
	140 mm thick	1.82	—	—	1.075	0.009	sq. m
	190 mm thick	2.02	—	—	1.075	0.013	sq. m
	215 mm thick	2.20	—	—	1.075	0.014	sq. m
F10.775	Walls; built curved; facing and pointing both sides; over 5.0 m radius on face						
	solid blocks						
	75 mm thick	1.41					
	100 mm thick	1.47	—	—	1.050	0.007	sq. m
	140 mm thick	1.72	—	—	1.050	0.009	sq. m
	190 mm thick	1.92	—	—	1.050	0.013	sq. m
	215 mm thick	2.11					
	hollow blocks						
	100 mm thick	1.41	—	—	1.075	0.007	sq. m
	140 mm thick	1.66	—	—	1.075	0.009	sq. m
	190 mm thick	1.84	—	—	1.075	0.013	sq. m
	215 mm thick	2.01	—	—	1.075	0.014	sq. m

		Labour (hr)			Blocks (sq. m)	Mortar (cu. m)	Unit
F10.780	Isolated piers, casings or chimney stacks; facing and pointing one side						
	solid blocks						
	75 mm thick	1.02	—	—	1.075	0.005	sq. m
	100 mm thick	1.07	—	—	1.050	0.007	sq. m
	140 mm thick	1.27	—	—	1.050	0.009	sq. m
	190 mm thick	1.42	—	—	1.050	0.013	sq. m
	215 mm thick	1.90	—	—	1.050	0.014	sq. m
	hollow blocks						
	100 mm thick	1.02	—	—	1.075	0.007	sq. m
	140 mm thick	1.23	—	—	1.075	0.009	sq. m
	190 mm thick	1.37	—	—	1.075	0.013	sq. m
	215 mm thick	1.49	—	—	1.075	0.014	sq. m
F10.785	Isolated piers, casings or chimney stacks; facing and pointing both sides						
	solid blocks						
	75 mm thick	1.16	—	—	1.075	0.005	sq. m
	100 mm thick	1.21	—	—	1.050	0.007	sq. m
	140 mm thick	1.41	—	—	1.050	0.009	sq. m
	190 mm thick	1.56	—	—	1.050	0.013	sq. m
	215 mm thick	2.04	—	—	1.050	0.014	sq. m
	hollow blocks						
	100 mm thick	1.16	—	—	1.075	0.007	sq. m
	140 mm thick	1.37	—	—	1.075	0.009	sq. m
	190 mm thick	1.51	—	—	1.075	0.013	sq. m
	215 mm thick	1.63	—	—	1.075	0.014	sq. m
F10.790	Projections; facing and pointing one side						
	solid blocks						
	215 wide × 100 mm thick	0.28	—	—	0.226	0.002	ln. m
	440 wide × 100 mm thick	0.56	—	—	0.462	0.003	ln. m
	215 wide × 140 mm thick	0.33	—	—	0.226	0.002	ln. m
	440 wide × 140 mm thick	0.67	—	—	0.462	0.004	ln. m
	215 wide × 190 mm thick	0.37	—	—	0.226	0.003	ln. m
	440 wide × 190 mm thick	0.76	—	—	0.462	0.006	ln. m
	hollow blocks						
	215 wide × 100 mm thick	0.26	—	—	0.226	0.002	ln. m
	440 wide × 100 mm thick	0.54	—	—	0.462	0.003	ln. m
	215 wide × 140 mm thick	0.32	—	—	0.226	0.002	ln. m
	440 wide × 140 mm thick	0.65	—	—	0.462	0.004	ln. m
	215 wide × 190 mm thick	0.35	—	—	0.226	0.003	ln. m
	440 wide × 190 mm thick	0.72	—	—	0.462	0.006	ln. m
	215 wide × 215 mm thick	0.39	—	—	0.226	0.003	ln. m
	440 wide × 215 mm thick	0.79	—	—	0.462	0.006	ln. m

	Labour (hr)			Blocks (sq. m)	Mortar (cu. m)	Unit
F10.805 **Precast concrete blocks, autoclaved, aerated; face size: 440 × 215 mm**						
F10.810 Walls; lightweight blocks						
75 mm thick	0.71	—	—	1.075	0.005	sq. m
100 mm thick	0.75	—	—	1.050	0.007	sq. m
115 mm thick	0.80	—	—	1.050	0.008	sq. m
125 mm thick	0.83	—	—	1.050	0.008	sq. m
130 mm thick	0.86	—	—	1.050	0.009	sq. m
135 mm thick	0.88	—	—	1.050	0.009	sq. m
150 mm thick	0.90	—	—	1.050	0.009	sq. m
200 mm thick	1.02	—	—	1.050	0.014	sq. m
F10.815 Walls; built curved; 2.5–5.0 m radius on face; lightweight blocks						
75 mm thick	1.19	—	—	1.075	0.005	sq. m
100 mm thick	1.25	—	—	1.050	0.007	sq. m
115 mm thick	1.34	—	—	1.050	0.008	sq. m
125 mm thick	1.39	—	—	1.050	0.008	sq. m
130 mm thick	1.44	—	—	1.050	0.009	sq. m
135 mm thick	1.47	—	—	1.050	0.009	sq. m
150 mm thick	1.50	—	—	1.050	0.009	sq. m
200 mm thick	1.70	—	—	1.050	0.014	sq. m
F10.820 Walls; built curved; over 5.0 m radius on face; lightweight blocks						
75 mm thick	1.07	—	—	1.075	0.005	sq. m
100 mm thick	1.13	—	—	1.050	0.007	sq. m
115 mm thick	1.20	—	—	1.050	0.008	sq. m
125 mm thick	1.25	—	—	1.050	0.008	sq. m
130 mm thick	1.29	—	—	1.050	0.009	sq. m
135 mm thick	1.32	—	—	1.050	0.009	sq. m
150 mm thick	1.35	—	—	1.050	0.009	sq. m
200 mm thick	1.53	—	—	1.050	0.014	sq. m
F10.825 Isolated piers, casings or chimney stacks; lightweight blocks						
75 mm thick	0.84	—	—	1.075	0.005	sq. m
100 mm thick	0.89	—	—	1.050	0.007	sq. m
115 mm thick	0.94	—	—	1.050	0.008	sq. m
125 mm thick	0.98	—	—	1.050	0.008	sq. m
130 mm thick	1.01	—	—	1.050	0.009	sq. m
135 mm thick	1.04	—	—	1.050	0.009	sq. m
150 mm thick	1.06	—	—	1.050	0.009	sq. m
200 mm thick	1.20	—	—	1.050	0.014	sq. m

		Labour (hr)			Blocks (sq. m)	Mortar (cu. m)	Unit
F10.830	Projections; lightweight blocks						
	215 wide × 100 mm thick	0.24	—	—	0.226	0.002	ln. m
	440 wide × 100 mm thick	0.49	—	—	0.462	0.003	ln. m
	215 wide × 115 mm thick	0.25	—	—	0.226	0.002	ln. m
	440 wide × 115 mm thick	0.52	—	—	0.462	0.003	ln. m
	215 wide × 125 mm thick	0.26	—	—	0.226	0.002	ln. m
	440 wide × 125 mm thick	0.54	—	—	0.462	0.004	ln. m
	215 wide × 130 mm thick	0.27	—	—	0.226	0.002	ln. m
	440 wide × 130 mm thick	0.56	—	—	0.462	0.004	ln. m
	215 wide × 135 mm thick	0.28	—	—	0.226	0.002	ln. m
	440 wide × 135 mm thick	0.57	—	—	0.462	0.004	ln. m
	215 wide × 150 mm thick	0.29	—	—	0.226	0.002	ln. m
	440 wide × 150 mm thick	0.58	—	—	0.462	0.005	ln. m
	215 wide × 200 mm thick	0.32	—	—	0.226	0.003	ln. m
	440 wide × 200 mm thick	0.66	—	—	0.462	0.006	ln. m
F10.835	Closing cavities with blockwork; 100 mm thick						
	at jambs or ends of walls	0.31	—	—	0.079	0.001	ln. m
	at sills	0.22	—	—	0.079	0.001	ln. m
	at tops of walls	0.16	—	—	0.079	0.001	ln. m
F10.840	Extra for bonding ends of blockwork to new brickwork; extra material for bonding; solid blocks						
	75 mm thick	—	—	—	0.054	0.001	ln. m
	100 mm thick	—	—	—	0.053	0.001	ln. m
	115 mm thick	—	—	—	0.053	0.001	ln. m
	125 mm thick	—	—	—	0.053	0.001	ln. m
	130 mm thick	—	—	—	0.053	0.001	ln. m
	135 mm thick	—	—	—	0.053	0.001	ln. m
	150 mm thick	—	—	—	0.053	0.001	ln. m
	200 mm thick	—	—	—	0.053	0.001	ln. m

	Labour (hr)			Blocks (sq. m)	Mortar (cu. m)	Unit

F10.845   **Precast concrete blocks, autoclaved, aerated; face size: 440 × 215 mm; fair face pointing as work proceeds**

F10.850   Walls; facing and pointing one side; lightweight blocks

	Labour (hr)			Blocks (sq. m)	Mortar (cu. m)	Unit
75 mm thick	0.85	—	—	1.075	0.005	sq. m
100 mm thick	0.89	—	—	1.050	0.007	sq. m
115 mm thick	0.94	—	—	1.050	0.008	sq. m
125 mm thick	0.97	—	—	1.050	0.008	sq. m
130 mm thick	1.00	—	—	1.050	0.009	sq. m
135 mm thick	1.02	—	—	1.050	0.009	sq. m
150 mm thick	1.04	—	—	1.050	0.009	sq. m
200 mm thick	1.18	—	—	1.050	0.014	sq. m

F10.855   Walls; facing and pointing both sides; lightweight blocks

	Labour (hr)			Blocks (sq. m)	Mortar (cu. m)	Unit
75 mm thick	0.99	—	—	1.075	0.005	sq. m
100 mm thick	1.03	—	—	1.050	0.007	sq. m
115 mm thick	1.08	—	—	1.050	0.008	sq. m
125 mm thick	1.11	—	—	1.050	0.008	sq. m
130 mm thick	1.14	—	—	1.050	0.009	sq. m
135 mm thick	1.16	—	—	1.050	0.009	sq. m
150 mm thick	1.18	—	—	1.050	0.009	sq. m
200 mm thick	1.32	—	—	1.050	0.014	sq. m

F10.860   Walls; built curved; facing and pointing one side; 2.5–5.0 m radius on face; lightweight blocks

	Labour (hr)			Blocks (sq. m)	Mortar (cu. m)	Unit
75 mm thick	1.33	—	—	1.075	0.005	sq. m
100 mm thick	1.39	—	—	1.050	0.007	sq. m
115 mm thick	1.48	—	—	1.050	0.008	sq. m
125 mm thick	1.53	—	—	1.050	0.008	sq. m
130 mm thick	1.58	—	—	1.050	0.009	sq. m
135 mm thick	1.61	—	—	1.050	0.009	sq. m
150 mm thick	1.64	—	—	1.050	0.009	sq. m
200 mm thick	1.84	—	—	1.050	0.014	sq. m

		Labour (hr)			Blocks (sq. m)	Mortar (cu. m)	Unit
F10.865	Walls; built curved; facing and pointing one side; over 5.0 m radius on face; lightweight blocks						
	75 mm thick	1.21	—	—	1.075	0.005	sq. m
	100 mm thick	1.27	—	—	1.050	0.007	sq. m
	115 mm thick	1.34	—	—	1.050	0.008	sq. m
	125 mm thick	1.39	—	—	1.050	0.008	sq. m
	130 mm thick	1.43	—	—	1.050	0.009	sq. m
	135 mm thick	1.46	—	—	1.050	0.009	sq. m
	150 mm thick	1.49	—	—	1.050	0.009	sq. m
	200 mm thick	1.67	—	—	1.050	0.014	sq. m
F10.870	Walls; built curved; facing and pointing both sides; 2.5–5.0 m radius on face; lightweight blocks						
	75 mm thick	1.47	—	—	1.075	0.005	sq. m
	100 mm thick	1.53	—	—	1.050	0.007	sq. m
	115 mm thick	1.62	—	—	1.050	0.008	sq. m
	125 mm thick	1.67	—	—	1.050	0.008	sq. m
	130 mm thick	1.72	—	—	1.050	0.009	sq. m
	135 mm thick	1.75	—	—	1.050	0.009	sq. m
	150 mm thick	1.78	—	—	1.050	0.009	sq. m
	200 mm thick	1.98	—	—	1.050	0.014	sq. m
F10.875	Walls; built curved; facing and pointing both sides; over 5.0 m radius on face; lightweight blocks						
	75 mm thick	1.35	—	—	1.075	0.005	sq. m
	100 mm thick	1.41	—	—	1.050	0.007	sq. m
	115 mm thick	1.48	—	—	1.050	0.008	sq. m
	125 mm thick	1.53	—	—	1.050	0.008	sq. m
	130 mm thick	1.57	—	—	1.050	0.009	sq. m
	135 mm thick	1.60	—	—	1.050	0.009	sq. m
	150 mm thick	1.63	—	—	1.050	0.009	sq. m
	200 mm thick	1.81	—	—	1.050	0.014	sq. m

	Labour (hr)			Blocks sq. m	Mortar (cu. m)	Unit
F10.880	Isolated piers, casings or chimney stacks; facing and pointing one side; lightweight blocks					
75 mm thick	0.98	—	—	1.075	0.005	sq. m
100 mm thick	1.03	—	—	1.050	0.007	sq. m
115 mm thick	1.08	—	—	1.050	0.008	sq. m
125 mm thick	1.12	—	—	1.050	0.008	sq. m
130 mm thick	1.15	—	—	1.050	0.009	sq. m
135 mm thick	1.18	—	—	1.050	0.009	sq. m
150 mm thick	1.20	—	—	1.050	0.009	sq. m
200 mm thick	1.34	—	—	1.050	0.014	sq. m
F10.885	Isolated piers, casings or chimney stacks; facing and pointing both sides; lightweight blocks					
75 mm thick	1.12	—	—	1.075	0.005	sq. m
100 mm thick	1.17	—	—	1.050	0.007	sq. m
115 mm thick	1.22	—	—	1.050	0.008	sq. m
125 mm thick	1.26	—	—	1.050	0.008	sq. m
130 mm thick	1.29	—	—	1.050	0.009	sq. m
135 mm thick	1.32	—	—	1.050	0.009	sq. m
150 mm thick	1.34	—	—	1.050	0.009	sq. m
200 mm thick	1.48	—	—	1.050	0.014	sq. m
F10.890	Projections; facing and pointing one side; lightweight blocks					
215 wide × 100 mm thick	0.27	—	—	0.226	0.002	ln. m
440 wide × 100 mm thick	0.55	—	—	0.462	0.003	ln. m
215 wide × 115 mm thick	0.28	—	—	0.226	0.002	ln. m
440 wide × 115 mm thick	0.58	—	—	0.462	0.003	ln. m
215 wide × 125 mm thick	0.29	—	—	0.226	0.002	ln. m
440 wide × 125 mm thick	0.60	—	—	0.462	0.004	ln. m
215 wide × 130 mm thick	0.30	—	—	0.226	0.002	ln. m
440 wide × 130 mm thick	0.62	—	—	0.462	0.004	ln. m
215 wide × 135 mm thick	0.31	—	—	0.226	0.002	ln. m
440 wide × 135 mm thick	0.63	—	—	0.462	0.004	ln. m
215 wide × 150 mm thick	0.32	—	—	0.226	0.002	ln. m
440 wide × 150 mm thick	0.64	—	—	0.462	0.005	ln. m
215 wide × 200 mm thick	0.35	—	—	0.226	0.003	ln. m
440 wide × 200 mm thick	0.72	—	—	0.462	0.006	ln. m

F30   *ACCESSORIES/SUNDRY ITEMS FOR BRICK/BLOCK WALLING*

F30.005   **Forming cavities**

F30.010   Forming cavities in hollow walls

F30.015   **Cavity wall insulation, wedging in position**

F30.020   Expanded polystyrene slabs

F30.025   Fibreglass batts

F30.030   **Damp-proof courses; polythene**

F30.035   Vertical

F30.040   Horizontal

F30.045   Horizontal; forming cavity gutters

F30.105   **Damp-proof courses; bitumen; BS.743, Table 1, Ref. A, B or C**

F30.110   Vertical

F30.115   Horizontal

F30.120   Horizontal; forming cavity gutters

F30.125   **Damp-proof courses; bitumen; BS.743, Table 1, Ref. F**

F30.130   Vertical

F30.135   Horizontal

F30.140   Horizontal; forming cavity gutters

F30.145	**Damp-proof courses; Hyload pitch polymer**
F30.150	Horizontal
F30.155	Vertical
F30.160	Horizontal; forming cavity gutters
F30.205	**Timloc stepped cavity trays**
F30.210	Type 'V' 300; 30, 50 and 75 mm wide cavity
F30.305	**Joint reinforcement**
F30.310	Exmet galvanized brickwork reinforcement; 24 gauge
F30.315	**Pointing in flashings**
F30.320	Raking out joints in brickwork for flashings; pointing
F30.325	Raking out joints in brickwork for asphalt skirting; pointing
F30.330	**Joints**
F30.335	Expansion joints in common brickwork; vertical
F30.340	Expansion joints in facing brickwork; vertical
F30.345	Expansion joints in blockwork; vertical
F30.405	**Flue linings**
F30.410	Flue linings, square section, rebated
F30.415	Flue linings, circular, rebated
F30.420	Fyrerite throat unit; red bank; refractory concrete
F30.425	Starters for flues
F30.430	Chimney pots

F30.435	**Air bricks**
F30.440	Air bricks; terracotta
F30.445	Cavity liners; terracotta
F30.450	**Gas flue blocks**
F30.455	Gas flue blocks; red bank RB2; buff fire-clay
F30.460	Gas flue blocks; red bank RB2; refractory concrete
F30.505	**Proprietary items**
F30.510	Fire-clay firebacks
F30.515	Cast iron bottom grate and vitreous enamel frets
F30.520	Open fire high output boiler unit
F30.525	Inset room heater
F30.530	Baxi Burnall underfloor draught fire with lift out ash box

F30   ACCESSORIES/SUNDRY ITEMS FOR BRICK/ BLOCK WALLING

		Labour (hr)	Wall ties (no.)	Insulation (sq. m)	Damp-proof course (sq. m)	Unit
F30.005	**Forming cavities**					
F30.010	Forming cavities in hollow walls; not exceeding 75 mm wide; BS.1243 wall ties; 4 per m² built in					
	200 mm butterfly wire ties	0.03	—	4.40	—	—	sq. m
	200 mm twisted steel ties	0.03	—	4.40	—	—	sq. m
	130 mm brick to timber ties	0.04	—	4.40	—	—	sq. m
F30.015	**Cavity wall insulation, wedging in position**					
F30.020	Expanded polystyrene slabs					
	25 mm thick	0.15	—	—	1.05	—	sq. m
	50 mm thick	0.15	—	—	1.05	—	sq. m
F30.025	Fibreglass batts					
	50 mm thick	0.15	—	—	1.05	—	sq. m
	75 mm thick	0.15	—	—	1.05	—	sq. m
F30.030	**Damp-proof courses; polythene damp-proof course; 100 mm laps; in cement mortar**					
F30.035	Vertical					
	not exceeding 225 mm wide	0.35	—	—	—	1.103	sq. m
	over 225 mm wide	0.33	—	—	—	1.103	sq. m
F30.040	Horizontal					
	not exceeding 225 mm wide	0.24	—	—	—	1.103	sq. m
	over 225 mm wide	0.23	—	—	—	1.103	sq. m
F30.045	Horizontal; forming cavity gutters in hollow walls					
	over 225 mm wide	0.46	—	—	—	1.103	sq. m

	Labour (hr)		Wall ties (no.)	Insulation (sq. m)	Damp-proof course (sq. m)	Unit
F30.105	**Damp-proof courses; bitumen damp-proof course, BS.743, Table 1, Ref. A, B, or C; 100 mm laps; in cement mortar**					
F30.110	Vertical					
	not exceeding 225 mm wide	0.35	—	—	—	1.103	sq. m
	over 225 mm wide	0.33	—	—	—	1.103	sq. m
F30.115	Horizontal					
	not exceeding 225 mm wide	0.24	—	—	—	1.103	sq. m
	over 225 mm wide	0.23	—	—	—	1.103	sq. m
F30.120	Horizontal, forming cavity gutters in hollow walls					
	over 225 mm wide	0.46	—	—	—	1.103	sq. m
F30.125	**Damp-proof courses; bitumen damp-proof course, BS.743, Table 1, Ref. F, asbestos and lead based; 100 mm laps; in cement mortar**					
F30.130	Horizontal					
	not exceeding 225 mm wide	0.37	—	—	—	1.103	sq. m
	over 225 mm wide	0.35	—	—	—	1.103	sq. m
F30.135	Vertical					
	not exceeding 225 mm wide	0.53	—	—	—	1.103	sq. m
	over 225 mm wide	0.50	—	—	—	1.103	sq. m
F30.140	Horizontal, forming cavity gutters in hollow walls					
	over 225 mm wide	0.75	—	—	—	1.103	sq. m
F30.145	**Damp-proof courses; Hyload pitch polymer damp-proof course; 100 mm laps; in cement mortar**					
F30.150	Horizontal					
	not exceeding 225 mm wide	0.26	—	—	—	1.103	sq. m
	over 225 mm wide	0.25	—	—	—	1.103	sq. m
F30.155	Vertical					
	not exceeding 225 mm wide	0.40	—	—	—	1.103	sq. m
	over 225 mm wide	0.38	—	—	—	1.103	sq. m
F30.160	Horizontal, forming cavity gutters in hollow walls					
	over 225 mm wide	0.56	—	—	—	1.103	sq. m

	Labour (hr)				Trays (no.)	Unit
F30.205	**Timloc stepped cavity trays**					
F30.210	Type 'V' 300 (measured along roof pitch angle), 30, 50 and 75 mm wide cavity					
20.5 degree	0.75	—	—	—	5.10	ln. m
22.5 degree	0.83	—	—	—	5.61	ln. m
25.0 degree	0.90	—	—	—	6.12	ln. m
27.5 degree	0.98	—	—	—	6.63	ln. m
30.0 degree	1.05	—	—	—	7.14	ln. m
32.5 degree	1.13	—	—	—	7.65	ln. m
35.0 degree	1.20	—	—	—	8.16	ln. m
37.5 degree	1.28	—	—	—	8.67	ln. m
40.0 degree	1.35	—	—	—	9.18	ln. m
42.5 degree	1.43	—	—	—	9.69	ln. m
45.0 degree	1.50	—	—	—	10.20	ln. m
47.5 degree	1.58	—	—	—	10.71	ln. m
50.0 degree	1.65	—	—	—	11.22	ln. m
straight through weep	0.18	—	—	—	1.02	no.
stop end weep	0.18	—	—	—	1.02	no.

		Labour (hr)	Exmet (ln. m)	Mastic (cc)	Flexcel (ln. m)	Cement Mortar (cu. m)	Unit
F30.305	**Joint reinforcement**						
F30.310	Exmet galvanized brickwork reinforcement; 24 gauge						
	65 mm wide	0.22	1.10	—	—	—	ln. m
	115 mm wide	0.28	1.10	—	—	—	ln. m
	175 mm wide	0.28	1.10	—	—	—	ln. m
F30.315	**Pointing in flashings**						
F30.320	Raking out joints in brickwork for turned in edges of flashings; pointing with cement mortar						
	horizontal	0.28	—	—	—	0.0001	ln. m
	stepped	0.28	—	—	—	0.0001	ln. m
F30.325	Raking out joints in brickwork and enlarging for nibs of asphalt skirting; pointing with cement mortar						
	horizontal	0.40	—	—	—	0.0001	ln. m
F30.330	**Joints**						
F30.335	Expansion joints in common brickwork; vertical; 10 mm thick Flexcel; pointing with polysulphide mastic						
	one side						
	102.5 mm wide	0.68	—	110	1.10	—	ln. m
	215.0 mm wide	0.87	—	110	1.10	—	ln. m
	both sides						
	102.5 mm wide	1.06	—	220	1.10	—	ln. m
	215.0 mm wide	1.25	—	220	1.10	—	ln. m
F30.340	Expansion joints in facing brickwork; vertical; 10 mm thick Flexcel; pointing with polysulphide mastic						
	one side						
	102.5 mm wide	0.68	—	110	1.10	—	ln. m
	215.0 mm wide	0.87	—	110	1.10	—	ln. m
	both sides						
	102.5 mm wide	1.06	—	220	1.10	—	ln. m
	225.0 mm wide	1.25	—	220	1.10	—	ln. m
F30.345	Expansion joints in blockwork; vertical; 10 mm thick Flexcel						
	100 mm wide	0.30	—	—	—	1.10	ln. m
	150 mm wide	0.40	—	—	—	1.10	ln. m

		Labour (hr)		Clayware (no.)	Mortar (cu. m)	High alumina cement (kg)	Unit
F30.405	**Flue linings**						
F30.410	Flue linings, square section, rebated; jointing in cement mortar; 185 × 185 mm internal size						
	300 mm long	0.30	—	3.50	0.005	—	In. m
	450 mm long	0.30	—	2.33	0.005	—	In. m
	600 mm long	0.30	—	1.75	0.005	—	In. m
	37.5 degree bend	0.60	—	1.05	0.002	—	no.
F30.415	Flue linings, circular, rebated; jointing in cement mortar; 185 mm internal diameter						
	300 mm long	0.30	—	3.50	0.016	—	In. m
	450 mm long	0.30	—	2.33	0.016	—	In. m
	600 mm long	0.30	—	1.75	0.016	—	In. m
	37.5 degree bend	0.60	—	1.05	0.006	—	no.
F30.420	Fyrerite Throat unit; Red Bank; refractory concrete; rebated joints in cement mortar						
	Throat unit; Ref: 740	2.00	—	1.05	0.005	—	no.
	Front brick; Ref: 744	0.22	—	1.05	—	—	no.
F30.425	Starters for nominal internal size flues						
	185 × 185 mm flue; Ref: 741	0.28	—	1.05	0.005	—	no.
	185 mm diameter flue; Ref: 742	0.28	—	1.05	0.005	—	no.
F30.430	Chimney pots; BS.1181; setting and flaunching in cement mortar; 185 mm internal diameter						
	300 mm long	0.38	—	1.025	0.020	—	no.
	450 mm long	0.38	—	1.025	0.020	—	no.
	600 mm long	0.38	—	1.025	0.020	—	no.
	750 mm long	0.38	—	1.025	0.020	—	no.
	900 mm long	0.38	—	1.025	0.020	—	no.

	Labour (hr)	Clayware (no.)	Mortar (cu. m)	Flue blocks (no.)	High alumina cement (kg)	Unit
F30.435 **Air bricks**						
F30.440 Air bricks, terra cotta; building into prepared openings						
220 × 70 mm	0.04	1.05	—	—	—	no.
220 × 145 mm	0.07	1.05	—	—	—	no.
220 × 220 mm	0.11	1.05	—	—	—	no.
F30.445 Cavity liners; terra cotta; building into prepared openings						
220 × 70 mm	0.04	—	1.05	—	—	no.
220 × 145 mm	0.07	—	1.05	—	—	no.
220 × 220 mm	0.11	—	1.05	—	—	no.
F30.450 **Gas flue blocks**						
F30.455 Gas flue blocks; Red Bank RB2 gas flue system; buff fire-clay; tapered rebate joints; jointing in high alumina cement mortar (1:4)						
flue blocks						
Ref. P2; 225 mm high	0.17	—	1.05	—	0.25	no.
Ref. R2; 300 mm high	0.19	—	1.05	—	0.25	no.
Ref. C2; 450 mm high	0.21	—	1.05	—	0.25	no.
flue blocks						
Ref. P; 225 mm high	0.15	—	1.05	—	0.25	no.
Ref. R; 300 mm high	0.17	—	1.05	—	0.25	no.
Ref. C; 450 mm high	0.19	—	1.05	—	0.25	no.
entry blocks						
Ref. K; 300 mm high	0.27	—	1.05	—	0.25	no.
Ref. KA; 300 mm high	0.25	—	1.05	—	0.25	no.
Ref. KB; 300 mm high	0.23	—	1.05	—	0.25	no.
offset blocks						
Ref. U; 235 mm high	0.19	—	1.05	—	0.25	no.
exit blocks						
Ref. Q; 300 mm high	0.27	—	1.05	—	0.25	no.
F30.460 Gas flue blocks; Red Bank RB2 gas flue system; refractory concrete; tapered rebate joints; jointing in high alumina cement mortar (1:4)						
recess block						
Ref. A7; 650 mm high	0.30	—	—	1.05	0.25	no.
cover block						
Ref. BB; 340 mm high	0.30	—	—	1.05	0.25	no.

		Labour (hr)	Stoves/ grates (no.)	Fire- clay (no.)	Clayware (no.)	Mortar (cu. m)	Unit
F30.505	**Proprietary items**						
F30.510	Fire-clay firebacks; bedding solidly in refactory mortar (1:4); 79 firebacks to BS.1251;						
	400 mm	1.14	—	1.00	—	0.020	no.
	450 mm	1.15	—	1.00	—	0.020	no.
F30.515	Cast iron bottom grate and vitreous enamel frets; setting in fireplace opening						
	400 mm	0.17	1.00	—	—	—	no.
	450 mm	0.17	1.00	—	—	—	no.
F30.520	Open fire high output boiler unit						
	400 mm	4.50	1.00	—	—	—	no.
	450 mm	5.00	1.00	—	—	—	no.
F30.525	Inset room heater appliances						
	non-boiler unit	3.50	1.00	—	—	—	no.
	with standard boiler	6.00	1.00	—	—	—	no.
	with high output boiler	6.00	1.00	—	—	—	no.
F30.530	'Baxi Burnall' underfloor draught fire with lift-out ash box and copper lustre or stainless steel front						
	400 mm	2.50	1.00	—	—	—	no.
	450 mm	2.50	1.00	—	—	—	no.

F31 *PRECAST CONCRETE SILLS/LINTELS/COPINGS*

F31.005	**Precast concrete sills**
F31.010	Sills, weathered and throated, 195 × 65 mm
F31.015	Sills, weathered and throated, 195 × 150 mm
F31.020	**Precast concrete lintels**
F31.025	Lintels; 103 × 150 mm plain rectangular section
F31.030	Lintels; 103 × 225 mm plain rectangular section
F31.035	Lintels; 150 to 103 × 150 mm splayed section
F31.040	Lintels; 150 to 103 × 225 mm splayed section
F31.050	**Precast concrete copings**
F31.055	Copings; once weathered; 915 mm long
F31.060	Copings; twice weathered; 915 mm long
F31.070	**Precast concrete cappings**
F31.075	Pier caps; weathered; 76 mm thick
F31.080	Chimney caps; weathered; 76 mm thick; once holed
F31.085	Chimney caps; weathered; 76 mm thick; twice holed

F31   PRECAST CONCRETE SILLS/LINTELS/COPINGS

		Labour (hr)			Units (no.)	Mortar (cu. m)	Unit
F31.005	**Precast concrete sills**						
F31.010	Sills; weathered and throated; bedded in mortar 195 × 65 mm;						
	600 mm long	0.11	—	—	1.025	0.001	no.
	900 mm long	0.17	—	—	1.025	0.001	no.
	1200 mm long	0.23	—	—	1.025	0.002	no.
	1500 mm long	0.25	—	—	1.025	0.002	no.
	1800 mm long	0.27	—	—	1.025	0.002	no.
	2100 mm long	0.29	—	—	1.025	0.003	no.
F31.015	Sills; weathered and throated; bedded in mortar 195 × 150 mm;						
	600 mm long	0.14	—	—	1.025	0.001	no.
	900 mm long	0.21	—	—	1.025	0.001	no.
	1200 mm long	0.28	—	—	1.025	0.002	no.
	1500 mm long	0.35	—	—	1.025	0.002	no.
	1800 mm long	0.41	—	—	1.025	0.002	no.
	2100 mm long	0.48	—	—	1.025	0.003	no.
F31.020	**Precast concrete lintels**						
F31.025	Lintels; 103 × 150 mm plain rectangular section; bedded in mortar						
	600 mm long	0.17	—	—	1.025	—	no.
	900 mm long	0.25	—	—	1.025	—	no.
	1200 mm long	0.33	—	—	1.025	—	no.
	1500 mm long	0.35	—	—	1.025	—	no.
	1800 mm long	0.42	—	—	1.025	—	no.
F31.030	Lintels; 103 × 225 mm plain rectangular section; bedding in gauged mortar						
	600 mm long	0.19	—	—	1.025	—	no.
	900 mm long	0.28	—	—	1.025	—	no.
	1200 mm long	0.37	—	—	1.025	—	no.
	1500 mm long	0.39	—	—	1.025	—	no.
	1800 mm long	0.47	—	—	1.025	—	no.
	2100 mm long	0.54	—	—	1.025	—	no.
	2400 mm long	0.62	—	—	1.025	—	no.
	2700 mm long	0.69	—	—	1.025	—	no.

		Labour (hr)			Units (no.)	Mortar (cu. m)	Unit
F31.035	Lintels; 150 to 103 × 150 mm splayed section; bedding in mortar						
	600 mm long	0.15	—	—	1.025	—	no.
	900 mm long	0.22	—	—	1.025	—	no.
	1200 mm long	0.30	—	—	1.025	—	no.
	1500 mm long	0.31	—	—	1.025	—	no.
	1800 mm long	0.38	—	—	1.025	—	no.
	2100 mm long	0.43	—	—	1.025	—	no.
	2400 mm long	0.50	—	—	1.025	—	no.
F31.040	Lintels; 150 to 103 × 225 mm splayed section; bedding in mortar						
	600 mm long	0.16	—	—	1.025	—	no.
	900 mm long	0.24	—	—	1.025	—	no.
	1200 mm long	0.33	—	—	1.025	—	no.
	1500 mm long	0.34	—	—	1.025	—	no.
	1800 mm long	0.41	—	—	1.025	—	no.
	2100 mm long	0.47	—	—	1.025	—	no.
	2400 mm long	0.54	—	—	1.025	—	no.
	2700 mm long	0.60	—	—	1.025	—	no.
	3000 mm long	0.64	—	—	1.025	—	no.
	3300 mm long	0.66	—	—	1.025	—	no.
	3600 mm long	0.72	—	—	1.025	—	no.
F31.050	**Precast concrete copings**						
F31.055	Copings; once weathered; bedding in mortar; 915 mm long						
	193 × 75 to 50 mm	0.23	—	—	1.025	0.001	no.
	305 × 75 to 50 mm	0.24	—	—	1.025	0.003	no.
	414 × 75 to 50 mm	0.27	—	—	1.025	0.004	no.
	fair end	0.10	—	—	—	—	no.
	square angle	0.26	—	—	—	—	no.
F31.060	Copings; twice weathered; bedding in mortar; 915 mm long						
	193 × 75 to 50 mm	0.20	—	—	1.025	0.001	no.
	305 × 75 to 50 mm	0.21	—	—	1.025	0.001	no.
	457 × 75 to 50 mm	0.23	—	—	1.025	0.003	no.
	fair end	0.10	—	—	—	—	no.
	square angle	0.26	—	—	—	—	no.

	Labour (hr)			Units (no.)	Mortar (cu. m)	Unit
F31.070	**Precast concrete cappings**					
F31.075	Pier caps; weathered; 76 mm thick					
305 × 305 mm	0.30	—	—	1.025	0.001	no.
457 × 457 mm	0.32	—	—	1.025	0.001	no.
533 × 533 mm	0.33	—	—	1.025	0.001	no.
F31.080	Chimney caps; weathered; 76 mm thick; once holed					
533 × 533 mm	0.38	—	—	1.025	0.002	no.
643 × 643 mm	0.40	—	—	1.025	0.002	no.
755 × 755 mm	0.43	—	—	1.025	0.003	no.
F31.085	Chimney caps; weathered; 76 mm thick; twice holed					
850 × 533 mm	0.50	—	—	1.025	0.003	no.

G STRUCTURAL/CARCASSING METAL/TIMBER

GA   **Notes**

1   The labour constants for work in this section allow for working with softwood; where hardwood is specified the labour constants should be increased by about 50 per cent.

2   The labour constants in Section G12 are based on a two and one bricklaying gang.

3   The labour constants in Section G20 are based on one craftsman.

G12   *ISOLATED STRUCTURAL METAL MEMBERS*

G12.005	**Weldable structural steel to BS.4360 in beams**
G12.010	Universal beams
G12.015	Joists
G12.105	**Lintels in galvanized steel**
G12.110	Type L1/S; I.G. lintels
G12.115	Type L1/S70; I.G. lintels
G12.120	Type L1/SW11; I.G. lintels
G12.125	Type L9; I.G. lintels
G12.130	Type L10; I.G. lintels
G12.135	Type box lintel 100; I.G. lintels
G12.140	Type internal; I.G. lintels

G12   ISOLATED STRUCTURAL METAL MEMBERS

		Labour (hr)			Lifting plant (hr)	Steel (tonne)	Unit

G12.005 | **Weldable structural steel to BS.4360 in beams, in single members cut to length, primed at works; ends built into structure**

G12.010 Universal beams
fixing at ground level

	Labour			Lifting	Steel	Unit
457 × 191 mm 82 kg/m	10.00	—	—	2.00	1.00	tonne
406 × 178 mm 60 kg/m	10.00	—	—	2.00	1.00	tonne
356 × 171 mm 51 kg/m	10.00	—	—	2.00	1.00	tonne
305 × 165 mm 46 kg/m	12.50	—	—	2.50	1.00	tonne
254 × 146 mm 37 kg/m	12.50	—	—	2.50	1.00	tonne

fixing 3.00 m above ground level

457 × 191 mm 82 kg/m	12.50	—	—	2.50	1.00	tonne
406 × 178 mm 60 kg/m	12.50	—	—	2.50	1.00	tonne
356 × 171 mm 51 kg/m	12.50	—	—	2.50	1.00	tonne
305 × 165 mm 46 kg/m	15.00	—	—	3.00	1.00	tonne
254 × 146 mm 37 kg/m	15.00	—	—	3.00	1.00	tonne

fixing 6.00 m above ground level

457 × 191 mm 82 kg/m	17.50	—	—	3.50	1.00	tonne
406 × 178 mm 60 kg/m	17.50	—	—	3.50	1.00	tonne
356 × 171 mm 51 kg/m	17.50	—	—	3.50	1.00	tonne
305 × 165 mm 46 kg/m	20.00	—	—	4.00	1.00	tonne
254 × 146 mm 37 kg/m	20.00	—	—	4.00	1.00	tonne

G12.015 Joists
fixing at ground level

254 × 114 mm 37.20 kg/m	12.50	—	—	2.50	1.00	tonne
203 × 102 mm 25.33 kg/m	12.50	—	—	2.50	1.00	tonne
178 × 102 mm 21.54 kg/m	12.50	—	—	2.50	1.00	tonne
152 × 76 mm 17.86 kg/m	17.50	—	—	3.50	1.00	tonne
127 × 76 mm 16.37 kg/m	17.50	—	—	3.50	1.00	tonne

fixing 3.00 m above ground level

254 × 114 mm 37.20 kg/m	15.00	—	—	3.00	1.00	tonne
203 × 102 mm 25.33 kg/m	15.00	—	—	3.00	1.00	tonne
178 × 102 mm 21.54 kg/m	15.00	—	—	3.00	1.00	tonne
152 × 76 mm 17.86 kg/m	20.00	—	—	4.00	1.00	tonne
127 × 76 mm 16.37 kg/m	20.00	—	—	4.00	1.00	tonne

fixing 6.00 m above ground level

254 × 114 mm 37.20 kg/m	20.00	—	—	4.00	1.00	tonne
203 × 102 mm 25.33 kg/m	20.00	—	—	4.00	1.00	tonne
178 × 102 mm 21.54 kg/m	20.00	—	—	4.00	1.00	tonne
152 × 76 mm 17.86 kg/m	25.00	—	—	5.00	1.00	tonne
127 × 76 mm 16.37 kg/m	25.00	—	—	5.00	1.00	tonne

	Labour (hr)				Lintel (nr)	Unit
G12.105	**Lintels in galvanized steel**					
G12.110	Type L1/S; I.G. lintels					
243 × 75 mm; length						
600 mm	0.09	—	—	—	1.00	no.
750 mm	0.10	—	—	—	1.00	no.
900 mm	0.10	—	—	—	1.00	no.
1050 mm	0.11	—	—	—	1.00	no.
1200 mm	0.13	—	—	—	1.00	no.
243 × 85 mm; length						
1350 mm	0.13	—	—	—	1.00	no.
1500 mm	0.14	—	—	—	1.00	no.
243 × 125 mm; length						
1650 mm	0.15	—	—	—	1.00	no.
1800 mm	0.17	—	—	—	1.00	no.
243 × 165 mm; length						
1950 mm	0.19	—	—	—	1.00	no.
2100 mm	0.21	—	—	—	1.00	no.
2250 mm	0.22	—	—	—	1.00	no.
2400 mm	0.23	—	—	—	1.00	no.
243 × 200 mm; length						
2550 mm	0.25	—	—	—	1.00	no.
2700 mm	0.27	—	—	—	1.00	no.
2850 mm	0.29	—	—	—	1.00	no.
3000 mm	0.30	—	—	—	1.00	no.
3150 mm	0.32	—	—	—	1.00	no.
3300 mm	0.33	—	—	—	1.00	no.
3450 mm	0.35	—	—	—	1.00	no.
3600 mm	0.38	—	—	—	1.00	no.
243 × 225 mm; length						
3750 mm	0.40	—	—	—	1.00	no.
3900 mm	0.43	—	—	—	1.00	no.
4050 mm	0.46	—	—	—	1.00	no.
4200 mm	0.50	—	—	—	1.00	no.
4350 mm	0.55	—	—	—	1.00	no.
243 × 230 mm; length						
4500 mm	0.60	—	—	—	1.00	no.
4650 mm	0.67	—	—	—	1.00	no.
4800 mm	0.75	—	—	—	1.00	no.

		Labour (hr)				Lintel (nr)	Unit
G12.115	Type L1/S70; I.G. lintels						
	253 × 80 mm; length						
	600 mm	0.10	—	—	—	1.00	no.
	750 mm	0.11	—	—	—	1.00	no.
	900 mm	0.11	—	—	—	1.00	no.
	1050 mm	0.13	—	—	—	1.00	no.
	1200 mm	0.14	—	—	—	1.00	no.
	253 × 100 mm; length						
	1350 mm	0.14	—	—	—	1.00	no.
	1500 mm	0.15	—	—	—	1.00	no.
	253 × 120 mm; length						
	1650 mm	0.17	—	—	—	1.00	no.
	1800 mm	0.19	—	—	—	1.00	no.
	253 × 160 mm; length						
	1950 mm	0.19	—	—	—	1.00	no.
	2100 mm	0.21	—	—	—	1.00	no.
	253 × 195 mm; length						
	2250 mm	0.22	—	—	—	1.00	no.
	2400 mm	0.23	—	—	—	1.00	no.
	2550 mm	0.25	—	—	—	1.00	no.
	2700 mm	0.27	—	—	—	1.00	no.
	2850 mm	0.29	—	—	—	1.00	no.
	3000 mm	0.30	—	—	—	1.00	no.
	3150 mm	0.32	—	—	—	1.00	no.
	3300 mm	0.33	—	—	—	1.00	no.
	3450 mm	0.35	—	—	—	1.00	no.
	3600 mm	0.38	—	—	—	1.00	no.
	253 × 220 mm; length						
	3750 mm	0.40	—	—	—	1.00	no.
	3900 mm	0.43	—	—	—	1.00	no.
	4050 mm	0.46	—	—	—	1.00	no.
	4200 mm	0.50	—	—	—	1.00	no.
	4350 mm	0.55	—	—	—	1.00	no.
	253 × 225 mm; length						
	4500 mm	0.60	—	—	—	1.00	no.
	4650 mm	0.67	—	—	—	1.00	no.
	4800 mm	0.75	—	—	—	1.00	no.

	Labour (hr)				Lintel (nr)	Unit
G12.120	Type L1/SWIL; I.G. lintels					
263 × 115 mm; length						
600 mm	0.12	—	—	—	1.00	no.
750 mm	0.12	—	—	—	1.00	no.
900 mm	0.13	—	—	—	1.00	no.
1050 mm	0.14	—	—	—	1.00	no.
1200 mm	0.16	—	—	—	1.00	no.
1350 mm	0.17	—	—	—	1.00	no.
1500 mm	0.18	—	—	—	1.00	no.
263 × 150 mm; length						
1650 mm	0.19	—	—	—	1.00	no.
1800 mm	0.21	—	—	—	1.00	no.
1950 mm	0.23	—	—	—	1.00	no.
2100 mm	0.27	—	—	—	1.00	no.
263 × 190 mm; length						
2250 mm	0.28	—	—	—	1.00	no.
2400 mm	0.30	—	—	—	1.00	no.
2550 mm	0.31	—	—	—	1.00	no.
2700 mm	0.33	—	—	—	1.00	no.
263 × 195 mm; length						
2850 mm	0.35	—	—	—	1.00	no.
3000 mm	0.37	—	—	—	1.00	no.
3150 mm	0.40	—	—	—	1.00	no.
3300 mm	0.43	—	—	—	1.00	no.
3450 mm	0.44	—	—	—	1.00	no.
3600 mm	0.46	—	—	—	1.00	no.
G12.125	Type L9; I.G. Lintels					
200 × 55 mm; length						
600 mm	0.08	—	—	—	1.00	no.
750 mm	0.08	—	—	—	1.00	no.
900 mm	0.09	—	—	—	1.00	no.
1050 mm	0.09	—	—	—	1.00	no.
1200 mm	0.11	—	—	—	1.00	no.
1350 mm	0.11	—	—	—	1.00	no.
1500 mm	0.12	—	—	—	1.00	no.
200 × 55 mm; length						
1650 mm	0.13	—	—	—	1.00	no.
1800 mm	0.15	—	—	—	1.00	no.
200 × 95 mm; length						
1950 mm	0.16	—	—	—	1.00	no.
2100 mm	0.19	—	—	—	1.00	no.
2250 mm	0.19	—	—	—	1.00	no.
2400 mm	0.21	—	—	—	1.00	no.
2550 mm	0.21	—	—	—	1.00	no.
2700 mm	0.23	—	—	—	1.00	no.

		Labour (hr)				Lintel (nr)	Unit
G12.130	Type L10; I.G. lintels						
	100 × 70 mm; length						
	600 mm	0.05	—	—	—	1.00	no.
	750 mm	0.05	—	—	—	1.00	no.
	900 mm	0.06	—	—	—	1.00	no.
	1050 mm	0.06	—	—	—	1.00	no.
	1200 mm	0.07	—	—	—	1.00	no.
	100 × 105 mm; length						
	1350 mm	0.08	—	—	—	1.00	no.
	1500 mm	0.08	—	—	—	1.00	no.
	1650 mm	0.09	—	—	—	1.00	no.
	1800 mm	0.10	—	—	—	1.00	no.
	100 × 205 mm; length						
	1950 mm	0.12	—	—	—	1.00	no.
	2100 mm	0.14	—	—	—	1.00	no.
	2250 mm	0.14	—	—	—	1.00	no.
	2400 mm	0.15	—	—	—	1.00	no.
	2550 mm	0.16	—	—	—	1.00	no.
	2700 mm	0.17	—	—	—	1.00	no.

	Labour (hr)				Lintel (nr)	Unit
G12.135 Type Box Lintel 100; I.G. lintels						
100 × 150 mm; length						
600 mm	0.09	—	—	—	1.00	no.
750 mm	0.10	—	—	—	1.00	no.
900 mm	0.10	—	—	—	1.00	no.
1050 mm	0.11	—	—	—	1.00	no.
1200 mm	0.13	—	—	—	1.00	no.
1350 mm	0.13	—	—	—	1.00	no.
1500 mm	0.13	—	—	—	1.00	no.
1650 mm	0.14	—	—	—	1.00	no.
1800 mm	0.15	—	—	—	1.00	no.
1950 mm	0.16	—	—	—	1.00	no.
2100 mm	0.17	—	—	—	1.00	no.
2250 mm	0.18	—	—	—	1.00	no.
2400 mm	0.18	—	—	—	1.00	no.
2550 mm	0.19	—	—	—	1.00	no.
2700 mm	0.20	—	—	—	1.00	no.
100 × 225 mm; length						
2850 mm	0.29	—	—	—	1.00	no.
3000 mm	0.30	—	—	—	1.00	no.
3150 mm	0.32	—	—	—	1.00	no.
3300 mm	0.33	—	—	—	1.00	no.
3450 mm	0.33	—	—	—	1.00	no.
3600 mm	0.38	—	—	—	1.00	no.
3750 mm	0.40	—	—	—	1.00	no.
3900 mm	0.43	—	—	—	1.00	no.
4050 mm	0.46	—	—	—	1.00	no.
4200 mm	0.50	—	—	—	1.00	no.
4350 mm	0.55	—	—	—	1.00	no.
4500 mm	0.60	—	—	—	1.00	no.
4650 mm	0.67	—	—	—	1.00	no.
4800 mm	0.75	—	—	—	1.00	no.
G12.140 Type Internal; I.G. Lintels						
64 × 28 mm; length						
1100 mm	0.09	—	—	—	1.00	no.
100 × 28 mm; length						
1100 mm	0.09	—	—	—	1.00	no.

G20   *CARPENTRY/TIMBER FRAMING/FIRST FIXING*

G20.005	**Trussed rafters, in softwood, impregnated**
G20.010	Fink pattern
G20.015	Monopitch pattern
G20.105	**Sawn softwood**
G20.110	Floor members
G20.115	Wall or partition members
G20.120	Plates
G20.125	Flat roof members
G20.130	Pitched roof members including ceiling joists
G20.135	Solid strutting to joists
G20.140	Herring-bone strutting
G20.145	Angle fillets
G20.150	Tilting fillets
G20.155	Rolls
G20.160	Firring pieces
G20.165	Bearers
G20.170	Bearers, framed
G20.175	Grounds and battens
G20.180	Open spaced grounds or battens; one way
G20.185	Open spaced grounds or battens; both ways

G20.205	**Wrought softwood**
G20.210	Boarding to gutter bottoms and sides
G20.215	Boarding fascias, barge boards or the like
G20.220	Boarding to eaves or the like; tongued and grooved boards
G20.305	**Plywood; BS.1455, 2/3 Grade; WBP bonded; butt joints**
G20.310	Boarding to gutter bottoms and sides
G20.315	Boarding to eaves, verges, fascias or the like
G20.320	Raking cutting on plywood boarding
G20.325	Curved cutting on plywood boarding
G20.405	**Non-asbestos boards to BS.476, Class 1; 'Supalux'**
G20.410	Boarding to eaves, verges, fascias or the like
G20.415	Raking cutting on 'Supalux' boarding
G20.420	Curved cutting on 'Supalux' boarding
G20.505	**Tempered hardboard; BS.1142, Part 2, Type TE; butt joints**
G20.510	Boarding to eaves, verges, fascias or the like
G20.515	Raking cutting on hardboard
G20.520	Curved cutting on hardboard

G20   CARPENTRY/TIMBER FRAMING/FIRST FIXING

		Labour (hr)				Trusses (no.)	Unit
G20.005	**Trussed rafters, in softwood, impregnated**						
G20.010	Fink pattern; 600 mm general spacing						
	22-30 degree pitch; span over plates						
	5000 mm	1.35	—	—	—	1.00	no.
	6000 mm	1.35	—	—	—	1.00	no.
	7000 mm	1.42	—	—	—	1.00	no.
	8000 mm	1.42	—	—	—	1.00	no.
	9000 mm	1.49	—	—	—	1.00	no.
	10000 mm	1.49	—	—	—	1.00	no.
	35-40 degree pitch; span over plates						
	5000 mm	1.42	—	—	—	1.00	no.
	6000 mm	1.42	—	—	—	1.00	no.
	7000 mm	1.49	—	—	—	1.00	no.
	8000 mm	1.49	—	—	—	1.00	no.
	9000 mm	1.56	—	—	—	1.00	no.
	10000 mm	1.56	—	—	—	1.00	no.
	45-50 degrees pitch; span over plates						
	5000 mm	1.49	—	—	—	1.00	no.
	6000 mm	1.49	—	—	—	1.00	no.
	7000 mm	1.56	—	—	—	1.00	no.
	8000 mm	1.95	—	—	—	1.00	no.
	9000 mm	1.95	—	—	—	1.00	no.
	10000 mm	2.15	—	—	—	1.00	no.
G20.015	Monopitch pattern; 600 mm general spacing						
	22-30 degree pitch; span over plates						
	2000 mm	0.69	—	—	—	1.00	no.
	3000 mm	0.69	—	—	—	1.00	no.
	4000 mm	0.72	—	—	—	1.00	no.
	5000 mm	0.72	—	—	—	1.00	no.
	6000 mm	0.83	—	—	—	1.00	no.
	35-40 degree pitch; span over plates						
	2000 mm	0.72	—	—	—	1.00	no.
	3000 mm	0.72	—	—	—	1.00	no.
	4000 mm	0.76	—	—	—	1.00	no.
	5000 mm	0.76	—	—	—	1.00	no.
	6000 mm	0.87	—	—	—	1.00	no.
	45-50 degree pitch; span over plates						
	2000 mm	0.76	—	—	—	1.00	no.
	3000 mm	0.76	—	—	—	1.00	no.
	4000 mm	0.80	—	—	—	1.00	no.
	5000 mm	0.80	—	—	—	1.00	no.
	6000 mm	0.91	—	—	—	1.00	no.

		Labour (hr)	Timber (ln. m)	Timber (cu. m)	Nails (kg)	Unit
G20.105	**Sawn softwood**					
G20.110	Floor members					
	50 × 75 mm	0.11	—	— 0.004031	0.004	ln. m
	50 × 100 mm	0.14	—	— 0.005375	0.005	ln. m
	50 × 125 mm	0.15	—	— 0.006719	0.007	ln. m
	50 × 150 mm	0.16	—	— 0.008063	0.008	ln. m
	50 × 175 mm	0.16	—	— 0.009406	0.009	ln. m
	50 × 200 mm	0.17	—	— 0.010750	0.011	ln. m
	50 × 225 mm	0.17	—	— 0.012094	0.012	ln. m
	75 × 150 mm	0.17	—	— 0.012094	0.012	ln. m
	75 × 175 mm	0.20	—	— 0.014109	0.014	ln. m
	75 × 200 mm	0.23	—	— 0.016125	0.016	ln. m
	75 × 225 mm	0.25	—	— 0.018141	0.018	ln. m
G20.115	Wall or partition members					
	38 × 50 mm	0.19	—	— 0.002043	0.017	ln. m
	38 × 75 mm	0.23	—	— 0.003064	0.025	ln. m
	38 × 100 mm	0.29	—	— 0.004085	0.033	ln. m
	50 × 50 mm	0.19	—	— 0.002688	0.022	ln. m
	50 × 75 mm	0.28	—	— 0.004031	0.033	ln. m
	50 × 100 mm	0.28	—	— 0.005375	0.044	ln. m
	75 × 75 mm	0.28	—	— 0.006047	0.050	ln. m
	75 × 100 mm	0.30	—	— 0.008063	0.066	ln. m
	100 × 100 mm	0.35	—	— 0.010750	0.088	ln. m
G20.120	Plates					
	25 × 75 mm	0.08	—	— 0.002016	0.002	ln. m
	25 × 100 mm	0.11	—	— 0.002688	0.003	ln. m
	25 × 125 mm	0.14	—	— 0.003359	0.003	ln. m
	25 × 150 mm	0.17	—	— 0.004031	0.004	ln. m
	38 × 100 mm	0.17	—	— 0.004085	0.004	ln. m
	38 × 125 mm	0.18	—	— 0.005106	0.005	ln. m
	38 × 150 mm	0.18	—	— 0.006128	0.006	ln. m
	50 × 50 mm	0.10	—	— 0.002688	0.003	ln. m
	50 × 75 mm	0.15	—	— 0.004031	0.004	ln. m
	50 × 100 mm	0.19	—	— 0.005375	0.005	ln. m
	50 × 125 mm	0.20	—	— 0.006719	0.007	ln. m
	50 × 150 mm	0.21	—	— 0.008063	0.008	ln. m

		Labour (hr)	Timber (ln. m)	Timber (cu. m)	Nails (kg)	Unit
G20.120	Plates (continued)					
	75 × 75 mm	0.18	—	— 0.006047	0.006	ln. m
	75 × 100 mm	0.21	—	— 0.008063	0.008	ln. m
	75 × 125 mm	0.23	—	— 0.010078	0.010	ln. m
	75 × 150 mm	0.23	—	— 0.012094	0.012	ln. m
G20.125	Flat roof members					
	50 × 75 mm	0.11	—	— 0.004031	0.004	ln. m
	50 × 100 mm	0.14	—	— 0.005375	0.005	ln. m
	50 × 125 mm	0.15	—	— 0.006719	0.007	ln. m
	50 × 150 mm	0.16	—	— 0.008063	0.008	ln. m
	50 × 175 mm	0.16	—	— 0.009406	0.009	ln. m
	50 × 200 mm	0.17	—	— 0.010750	0.011	ln. m
	50 × 225 mm	0.17	—	— 0.012094	0.012	ln. m
	75 × 150 mm	0.17	—	— 0.012094	0.012	ln. m
	75 × 175 mm	0.20	—	— 0.014109	0.014	ln. m
	75 × 200 mm	0.23	—	— 0.016125	0.016	ln. m
	75 × 225 mm	0.25	—	— 0.018141	0.018	ln. m
G20.130	Pitched roof members including ceiling joists					
	25 × 100 mm	0.14	—	— 0.002688	0.009	ln. m
	25 × 125 mm	0.17	—	— 0.003359	0.011	ln. m
	25 × 150 mm	0.21	—	— 0.004031	0.013	ln. m
	25 × 175 mm	0.24	—	— 0.004703	0.015	ln. m
	50 × 75 mm	0.18	—	— 0.004031	0.013	ln. m
	50 × 100 mm	0.23	—	— 0.005375	0.017	ln. m
	50 × 125 mm	0.24	—	— 0.006719	0.022	ln. m
	50 × 150 mm	0.25	—	— 0.008063	0.026	ln. m
	50 × 175 mm	0.25	—	— 0.009406	0.030	ln. m
	50 × 200 mm	0.27	—	— 0.010750	0.034	ln. m
	50 × 225 mm	0.27	—	— 0.012094	0.039	ln. m
	75 × 100 mm	0.25	—	— 0.008063	0.026	ln. m
	75 × 125 mm	0.27	—	— 0.010078	0.032	ln. m
	75 × 150 mm	0.27	—	— 0.012094	0.039	ln. m
	75 × 175 mm	0.32	—	— 0.014109	0.045	ln. m
	75 × 200 mm	0.36	—	— 0.016125	0.052	ln. m
	75 × 225 mm	0.41	—	— 0.018141	0.058	ln. m

		Labour (hr)		Timber (ln. m)	Timber (cu. m)	Nails (kg)	Unit
G20.135	Solid strutting to joists						
	50 × 100 mm	0.37	—	—	0.005375	0.006	ln. m
	50 × 150 mm	0.37	—	—	0.008063	0.091	ln. m
	50 × 175 mm	0.37	—	—	0.009406	0.106	ln. m
	50 × 200 mm	0.38	—	—	0.010750	0.121	ln. m
	50 × 225 mm	0.38	—	—	0.012094	0.137	ln. m
G20.140	Herring-bone strutting						
	38 × 38 mm to joists						
	150 mm deep	0.38	—	—	0.002708	0.061	ln. m
	175 mm deep	0.38	—	—	0.002708	0.061	ln. m
	200 mm deep	0.38	—	—	0.002888	0.065	ln. m
	225 mm deep	0.38	—	—	0.003249	0.073	ln. m
	50 × 50 mm to joists						
	150 mm deep	0.45	—	—	0.004688	0.061	ln. m
	175 mm deep	0.45	—	—	0.004688	0.061	ln. m
	200 mm deep	0.45	—	—	0.005000	0.065	ln. m
	225 mm deep	0.45	—	—	0.005625	0.073	ln. m
G20.145	Angle fillets						
	38 × 38 mm	0.17	—	1.075	—	0.02	ln. m
	50 × 50 mm	0.18	—	1.075	—	0.02	ln. m
	75 × 75 mm	0.20	—	1.075	—	0.02	ln. m
G20.150	Tilting fillets;						
	25 × 50 mm	0.10	—	1.075	—	0.02	ln. m
	38 × 75 mm	0.13	—	1.075	—	0.02	ln. m
G20.155	Rolls;						
	50 × 50 mm	0.21	—	1.075	—	0.02	ln. m
	50 × 75 mm	0.22	—	1.075	—	0.02	ln. m
G20.160	Firring pieces;						
	50 mm wide						
	average depth; 25 mm	0.02	—	1.075	—	0.02	ln. m
	50 mm	0.04	—	1.075	—	0.02	ln. m
	63 mm	0.05	—	1.075	—	0.02	ln. m
	75 mm	0.07	—	1.075	—	0.02	ln. m
	75 mm wide						
	average depth; 25 mm	0.03	—	1.075	—	0.02	ln. m
	50 mm	0.07	—	1.075	—	0.02	ln. m
	63 mm	0.08	—	1.075	—	0.02	ln. m
	75 mm	0.10	—	1.075	—	0.02	ln. m

		Labour (hr)			Timber (ln. m)	Nails (kg)	Unit
G20.165	Bearers						
	25 × 38 mm	0.13	—	—	1.075	0.02	ln. m
	25 × 50 mm	0.14	—	—	1.075	0.02	ln. m
	38 × 38 mm	0.14	—	—	1.075	0.02	ln. m
	38 × 50 mm	0.15	—	—	1.075	0.02	ln. m
G20.170	Bearers, framed						
	25 × 38 mm	0.27	—	—	1.075	0.03	ln. m
	25 × 50 mm	0.29	—	—	1.075	0.03	ln. m
	38 × 38 mm	0.29	—	—	1.075	0.03	ln. m
	38 × 50 mm	0.31	—	—	1.075	0.03	ln. m
G20.175	Grounds and battens						
	25 × 38 mm	0.15	—	—	1.075	0.02	ln. m
	25 × 50 mm	0.15	—	—	1.075	0.02	ln. m
	38 × 50 mm	0.17	—	—	1.075	0.02	ln. m
G20.180	Open spaced grounds or battens to receive boarded finish; one way						
	25 × 38 mm at 400 mm centres	0.38	—	—	2.690	0.04	sq. m
	25 × 50 mm at 400 mm centres	0.38	—	—	2.690	0.04	sq. m
	38 × 50 mm at 400 mm centres	0.43	—	—	2.690	0.04	sq. m
	25 × 38 mm at 450 mm centres	0.33	—	—	2.390	0.04	sq. m
	25 × 50 mm at 450 mm centres	0.33	—	—	2.390	0.04	sq. m
	38 × 50 mm at 450 mm centres	0.37	—	—	2.390	0.04	sq. m
	25 × 38 mm at 600 mm centres	0.25	—	—	1.790	0.03	sq. m
	25 × 50 mm at 600 mm centres	0.25	—	—	1.790	0.03	sq. m
	38 × 50 mm at 600 mm centres	0.28	—	—	1.790	0.03	sq. m
G20.185	Open spaced grounds or battens to receive boarded finish; both ways						
	25 × 38 mm at 400 mm centres	0.98	—	—	6.181	0.09	sq. m
	25 × 50 mm at 400 mm centres	0.98	—	—	6.181	0.09	sq. m
	38 × 50 mm at 400 mm centres	1.04	—	—	6.181	0.09	sq. m
	25 × 38 mm at 450 mm centres	0.88	—	—	5.590	0.08	sq. m
	25 × 50 mm at 450 mm centres	0.88	—	—	5.590	0.08	sq. m
	38 × 50 mm at 450 mm centres	0.94	—	—	5.590	0.08	sq. m
	25 × 38 mm at 600 mm centres	0.69	—	—	4.386	0.06	sq. m
	25 × 50 mm at 600 mm centres	0.69	—	—	4.386	0.06	sq. m
	38 × 50 mm at 600 mm centres	0.73	—	—	4.386	0.06	sq. m

	Labour (hr)			Timber (ln. m)	Nails (kg)	Unit
G20.205	**Wrought softwood**					
G20.210	Boarding to gutter bottoms and sides; 125 mm nominal width boards; tongued and grooved joints					
	19 mm thick	2.94	—	—	9.43	0.32	sq. m
	25 mm thick	2.94	—	—	9.43	0.32	sq. m
G20.215	Boarding to fascias, barge boards or the like; 25 mm thick					
	150 mm wide	0.33	—	—	1.08	0.04	ln. m
	175 mm wide	0.33	—	—	1.08	0.06	ln. m
	200 mm wide	0.33	—	—	1.08	0.08	ln. m
	225 mm wide	0.35	—	—	1.08	0.08	ln. m
G20.220	Boarding to eaves, verges fascias or the like; 100 mm nominal width boards; tongued, grooved and V-jointed one side matchboarding					
	13 mm thick					
	150 mm wide	0.43	—	—	2.15	0.06	ln. m
	200 mm wide	0.43	—	—	3.23	0.08	ln. m
	225 mm wide	0.49	—	—	3.23	0.08	ln. m
	250 mm wide	0.65	—	—	3.23	0.08	ln. m
	over 300 mm wide	1.72	—	—	12.04	0.40	sq. m
	19 mm thick					
	150 mm wide	0.43	—	—	2.15	0.06	ln. m
	200 mm wide	0.43	—	—	3.23	0.08	ln. m
	225 mm wide	0.49	—	—	3.23	0.08	ln. m
	250 mm wide	0.65	—	—	3.23	0.08	ln. m
	over 300 mm wide	1.72	—	—	12.04	0.40	sq. m

	Labour (hr)			Boarding (sq. m)	Nails (kg)	Unit
G20.305	**Plywood; BS.1455, 2/3 grade; WBP bonded; butt joints**					
G20.310	Boarding to gutter bottoms and sides					
	150 mm wide					
	6 mm thick	0.50	—	—	0.17	0.01	ln. m
	9 mm thick	0.56	—	—	0.17	0.01	ln. m
	12 mm thick	0.57	—	—	0.17	0.01	ln. m
	18 mm thick	0.63	—	—	0.17	0.01	sq. m
	25 mm thick	0.78	—	—	0.17	0.01	ln. m
	225 mm wide					
	6 mm thick	0.63	—	—	0.26	0.02	ln. m
	9 mm thick	0.69	—	—	0.26	0.02	ln. m
	12 mm thick	0.71	—	—	0.26	0.02	ln. m
	18 mm thick	0.78	—	—	0.26	0.02	ln. m
	25 mm thick	0.80	—	—	0.26	0.02	ln. m
G20.315	Boarding to eaves, verges, fascias or the like					
	6 mm thick					
	150 mm wide	0.24	—	—	0.17	0.02	ln. m
	175 mm wide	0.24	—	—	0.20	0.02	ln. m
	200 mm wide	0.25	—	—	0.23	0.03	ln. m
	225 mm wide	0.29	—	—	0.26	0.03	ln. m
	250 mm wide	0.30	—	—	0.29	0.04	ln. m
	300 mm wide	0.32	—	—	0.35	0.04	ln. m
	over 300 mm wide	1.05	—	—	1.15	0.06	sq. m
	9 mm thick					
	150 mm wide	0.25	—	—	0.17	0.02	ln. m
	175 mm wide	0.25	—	—	0.20	0.02	ln. m
	200 mm wide	0.26	—	—	0.23	0.03	ln. m
	225 mm wide	0.30	—	—	0.26	0.03	ln. m
	250 mm wide	0.32	—	—	0.29	0.04	ln. m
	300 mm wide	0.33	—	—	0.35	0.04	ln. m
	over 300 mm wide	1.11	—	—	1.15	0.06	sq. m
	12 mm thick					
	150 mm wide	0.28	—	—	0.17	0.02	ln. m
	175 mm wide	0.28	—	—	0.20	0.02	ln. m
	200 mm wide	0.29	—	—	0.23	0.03	ln. m
	225 mm wide	0.33	—	—	0.26	0.03	ln. m
	250 mm wide	0.35	—	—	0.29	0.04	ln. m
	300 mm wide	0.37	—	—	0.35	0.04	ln. m
	over 300 mm wide	1.25	—	—	1.15	0.06	sq. m

		Labour (hr)			Boarding (sq. m)	Nails (kg)	Unit
G20.315	Boarding to eaves, verges, fascias or the like (continued)						
	18 mm thick						
	150 mm wide	0.29	—	—	0.17	0.02	ln. m
	175 mm wide	0.29	—	—	0.20	0.02	ln. m
	200 mm wide	0.31	—	—	0.23	0.03	ln. m
	225 mm wide	0.32	—	—	0.26	0.03	ln. m
	250 mm wide	0.33	—	—	0.29	0.04	ln. m
	300 mm wide	0.35	—	—	0.35	0.04	ln. m
	over 300 mm wide	1.33	—	—	1.15	0.06	sq. m
	25 mm thick						
	150 mm wide	0.31	—	—	0.17	0.02	ln. m
	175 mm wide	0.31	—	—	0.20	0.02	ln. m
	200 mm wide	0.29	—	—	0.23	0.03	ln. m
	225 mm wide	0.33	—	—	0.26	0.03	ln. m
	250 mm wide	0.35	—	—	0.29	0.04	ln. m
	300 mm wide	0.37	—	—	0.35	0.04	ln. m
	over 300 mm wide	1.43	—	—	1.15	0.06	sq. m
G20.320	Raking cutting on plywood boarding						
	6 mm thick	0.25	—	—	—	—	ln. m
	9 mm thick	0.29	—	—	—	—	ln. m
	12 mm thick	0.32	—	—	—	—	ln. m
	18 mm thick	0.35	—	—	—	—	ln. m
	25 mm thick	0.37	—	—	—	—	ln. m
G20.325	Curved cutting on plywood boarding						
	6 mm thick	0.50	—	—	—	—	ln. m
	9 mm thick	0.58	—	—	—	—	ln. m
	12 mm thick	0.64	—	—	—	—	ln. m
	18 mm thick	0.70	—	—	—	—	ln. m
	25 mm thick	0.74	—	—	—	—	ln. m

		Labour (hr)			Boarding (sq. m)	Nails (kg)	Unit
G20.405	**Non-asbestos boards to BS 476, flameproof to class 1; 'Supalux'**						
G20.410	Boarding to eaves, verges, fascias or the like						
	6 mm thick						
	150 mm wide	0.42	—	—	0.17	0.01	ln. m
	175 mm wide	0.43	—	—	0.20	0.01	ln. m
	200 mm wide	0.44	—	—	0.23	0.01	ln. m
	225 mm wide	0.50	—	—	0.26	0.02	ln. m
	250 mm wide	0.53	—	—	0.29	0.02	ln. m
	300 mm wide	0.56	—	—	0.35	0.03	ln. m
	over 300 mm wide	1.67	—	—	1.15	0.04	sq. m
	9 mm thick						
	150 mm wide	0.43	—	—	0.17	0.01	ln. m
	175 mm wide	0.44	—	—	0.20	0.01	ln. m
	200 mm wide	0.50	—	—	0.23	0.01	ln. m
	225 mm wide	0.53	—	—	0.26	0.02	ln. m
	250 mm wide	0.56	—	—	0.29	0.02	ln. m
	300 mm wide	0.59	—	—	0.35	0.03	ln. m
	over 300 mm wide	1.82	—	—	1.15	0.04	sq. m
	12 mm thick						
	150 mm wide	0.44	—	—	0.17	0.01	ln. m
	175 mm wide	0.50	—	—	0.20	0.01	ln. m
	200 mm wide	0.53	—	—	0.23	0.01	ln. m
	225 mm wide	0.56	—	—	0.26	0.02	ln. m
	250 mm wide	0.59	—	—	0.29	0.02	ln. m
	300 mm wide	0.63	—	—	0.35	0.03	ln. m
	over 300 mm wide	2.00	—	—	1.15	0.04	sq. m
G20.415	Raking cutting on 'Supalux' boarding						
	6 mm thick	0.29	—	—	0.15	—	ln. m
	9 mm thick	0.29	—	—	0.15	—	ln. m
	12 mm thick	0.29	—	—	0.15	—	ln. m
G20.420	Curved cutting on 'Supalux' boarding						
	6 mm thick	0.58	—	—	0.15	—	ln. m
	9 mm thick	0.58	—	—	0.15	—	ln. m
	12 mm thick	0.58	—	—	0.15	—	ln. m

	Labour (hr)			Boarding (sq. m)	Nails (kg)	Unit
G20.505	**Tempered hardboard; BS.1142, Part 2, Type TE; butt joints**					
G20.510	Boarding to eaves, verges, fascias or the like					
	3.2 mm thick					
	150 wide	0.24	—	—	0.17	0.02	In. m
	175 wide	0.24	—	—	0.20	0.02	In. m
	200 wide	0.25	—	—	0.23	0.03	In. m
	225 wide	0.29	—	—	0.26	0.03	In. m
	250 wide	0.30	—	—	0.29	0.03	In. m
	300 wide	0.32	—	—	0.35	0.04	In. m
	over 300 wide	1.05	—	—	1.15	0.06	sq. m
	6.4 mm thick					
	150 wide	0.24	—	—	0.17	0.02	In. m
	175 wide	0.24	—	—	0.20	0.02	In. m
	200 wide	0.25	—	—	0.23	0.03	In. m
	225 wide	0.29	—	—	0.26	0.03	In. m
	250 wide	0.30	—	—	0.29	0.04	In. m
	300 wide	0.32	—	—	0.35	0.04	In. m
	over 300 wide	1.05	—	—	1.15	0.06	sq. m
G20.515	Raking cutting on hardboard;					
	3.2 mm thick	0.12	—	—	—	—	In. m
	6.4 mm thick	0.12	—	—	—	—	In. m
G20.520	Curved cutting on hardboard					
	3.2 mm thick	0.25	—	—	—	—	In. m
	6.4 mm thick	0.25	—	—	—	—	In. m

H CLADDING/COVERING

HA	**Notes**

1   The labour constants in Section H21 are based on one craftsman joiner.

2   The labour constants in Section H60 are based on one craftsman roof tiler and one labourer.

3   Work to dormers or the like add 50 per cent to labour constants.

H21   *TIMBER WEATHERBOARDING*

H21.005   **Wrought softwood**

H21.010   Boarding to walls; shiplap weatherboarding; horizontal

H21.015   Boarding to walls; shiplap weatherboarding; diagonally

H21.020   Raking cutting on boarding

H21.025   Curved cutting on boarding

H21.030   Notches; per 25 mm girth on boarding

H21.035   Cutting and fitting around obstructions; per 25 mm girth

H21 TIMBER WEATHERBOARDING

		Labour (hr)			Timber (ln. m)	Nails (kg)	Unit
H21.005	**Wrought softwood**						
H21.010	Boarding to walls; shiplap weatherboarding; tongued and grooved joints; horizontal						
	100 mm nominal width boards						
	13 mm thick	1.18	—	—	12.04	0.40	sq. m
	19 mm thick	1.18	—	—	12.04	0.40	sq. m
	25 mm thick	1.18	—	—	12.04	0.40	sq. m
	125 mm nominal width boards						
	13 mm thick	0.93	—	—	9.43	0.32	sq. m
	19 mm thick	0.93	—	—	9.43	0.32	sq. m
	25 mm thick	0.93	—	—	9.43	0.32	sq. m
H21.015	Boarding to walls; shiplap weatherboarding; tongued and grooved joints; diagonally						
	100 mm nominal width boards						
	13 mm thick	1.39	—	—	12.04	0.40	sq. m
	19 mm thick	1.39	—	—	12.04	0.40	sq. m
	25 mm thick	1.39	—	—	12.04	0.40	sq. m
	125 mm nominal width boards						
	13 mm thick	1.10	—	—	9.43	0.32	sq. m
	19 mm thick	1.10	—	—	9.43	0.32	sq. m
	25 mm thick	1.10	—	—	9.43	0.32	sq. m
H21.020	Raking cutting on boarding						
	13 mm thick	0.13	—	—	—	—	ln. m
	19 mm thick	0.13	—	—	—	—	ln. m
	25 mm thick	0.13	—	—	—	—	ln. m
H21.025	Curved cutting on boarding						
	13 mm thick	0.29	—	—	—	—	ln. m
	19 mm thick	0.29	—	—	—	—	ln. m
	25 mm thick	0.29	—	—	—	—	ln. m
H21.030	Notches; per 25 mm girth on boarding						
	13 mm thick	0.17	—	—	—	—	no.
	19 mm thick	0.20	—	—	—	—	no.
	25 mm thick	0.25	—	—	—	—	no.
H21.035	Cutting and fitting around obstructions; per 25 mm girth on boarding						
	13 mm thick	0.20	—	—	—	—	no.
	19 mm thick	0.25	—	—	—	—	no.
	25 mm thick	0.30	—	—	—	—	no.

H60 *CLAY/CONCRETE ROOF TILING*

H60.005	**Concrete interlocking tiles**
H60.010	Roof coverings; members at 400 mm centres
H60.015	Roof coverings; members at 450 mm centres
H60.020	Roof coverings; members at 600 mm centres
H60.025	Square cutting
H60.030	Raking cutting
H60.035	Curved cutting
H60.040	**Extra over pitched roof coverings**
H60.045	Fixing every tile with one aluminium nail
H60.050	Fixing every tile at eaves with one aluminium nail
H60.055	Verges; 6 mm asbestos undercloak, bedding and pointing in cement mortar
H60.060	Verges; treble roll left-hand verge tile; 6 mm asbestos undercloak
H60.065	Ridge tiles; half-round; bedding and pointing in cement mortar
H60.070	Ends filled with mortar
H60.075	Mitred angles
H60.080	Mitred three-way intersection

H60.105	**Clayware tiles; Sandtoft Goxhill; on 25 × 38 mm battens**
H60.110	Roof coverings; members at 400 mm centres
H60.115	Roof coverings; members at 450 mm centres
H60.120	Roof coverings; members at 600 mm centres
H60.125	Square cutting at abutment of large opening
H60.130	Square cutting at top edges of large opening
H60.135	Raking cutting
H60.140	Curved cutting
H60.150	Cutting to valleys
H60.205	**Extra over pitched roof coverings for**
H60.210	Fixing each tile in every alternative course with aluminium alloy nail
H60.215	Fixing every tile with aluminium alloy nail
H60.220	Extra undercloak plain tiles at eaves
H60.225	Verges; single extra undercloak coarse plain tiles
H60.230	Verges; double roll tiles; single extra undercloak course plain tiles
H60.235	Ridge tiles; half-round; double dentil slips both sides
H60.240	Extra over ridge tiles

H60 CLAY/CONCRETE ROOF TILING

	Labour (hr)	Tiles (no.)	Battens (ln. m)	Round wire nails (kg)	Roofing felt (sq. m)	Unit
H60.005 **Roofing; concrete interlocking tiles; BS.747 type IF reinforced bitumen felt underlay; 150 mm laps; secured with softwood battens, impregnated, fixed with 65 × 3.35 mm round wire nails**						
H60.010 Roof coverings; members at 400 mm centres						
75 mm headlap; 19 × 38 mm battens						
430 × 380 mm tiles	0.36	8.62	2.99	0.036	1.15	sq. m
418 × 330 mm tiles	0.38	10.19	3.06	0.037	1.15	sq. m
381 × 227 mm tiles	0.52	16.70	3.42	0.042	1.15	sq. m
100 mm headlap; 19 × 38 mm battens						
430 × 380 mm tiles	0.39	9.30	3.23	0.039	1.15	sq. m
418 × 330 mm tiles	0.41	11.05	3.32	0.040	1.15	sq. m
381 × 227 mm tiles	0.57	18.25	3.74	0.046	1.15	sq. m
H60.015 Roof coverings; members at 450 mm centres						
75 mm headlap; 19 × 38 mm battens						
430 × 380 mm tiles	0.36	8.62	2.99	0.032	1.15	sq. m
418 × 330 mm tiles	0.38	10.19	3.06	0.033	1.15	sq. m
381 × 227 mm tiles	0.52	16.70	3.42	0.037	1.15	sq. m
100 mm headlap; 19 × 38 mm battens						
430 × 380 mm tiles	0.39	9.30	3.23	0.034	1.15	sq. m
418 × 330 mm tiles	0.41	11.05	3.32	0.036	1.15	sq. m
381 × 227 mm tiles	0.57	18.25	3.74	0.040	1.15	sq. m
H60.020 Roof coverings; members at 600 mm centres						
75 mm headlap; 25 × 38 mm battens						
430 × 380 mm tiles	0.36	8.62	2.99	0.024	1.15	sq. m
418 × 330 mm tiles	0.38	10.19	3.06	0.025	1.15	sq. m
381 × 227 mm tiles	0.52	16.70	3.42	0.028	1.15	sq. m
100 mm headlap; 25 × 38 mm battens						
430 × 380 mm tiles	0.39	9.30	3.23	0.026	1.15	sq. m
418 × 330 mm tiles	0.41	11.05	3.32	0.027	1.15	sq. m
381 × 227 mm tiles	0.57	18.25	3.74	0.030	1.15	sq. m
H60.025 Square cutting;						
at abutment of large opening	0.20	0.65	—	—	—	ln. m
at top edges of large openings	0.32	0.65	—	—	—	ln. m
H60.030 Raking cutting	0.38	1.03	—	—	—	ln. m
H60.035 Curved cutting	0.46	1.23	—	—	—	ln. m

		Labour (hr)	Tiles (no.)	Asbestos undercloak (ln. m)	Cement mortar (cu. m)	Aluminium nails (kg)	Unit
H60.040	**Extra over pitched roof coverings for**						
H60.045	Fixing every tile with one 65 × 3.75 mm aluminium alloy nail						
	75 mm lap						
	430 × 380 mm tiles	0.07	—	—	—	0.018	sq. m
	418 × 330 mm tiles	0.08	—	—	—	0.022	sq. m
	381 × 227 mm tiles	0.10	—	—	—	0.035	sq. m
	100 mm lap						
	430 × 380 mm tiles	0.07	—	—	—	0.021	sq. m
	418 × 330 mm tiles	0.08	—	—	—	0.025	sq. m
	381 × 227 mm tiles	0.10	—	—	—	0.040	sq. m
H60.050	Fixing every tile with one 65 × 3.75 mm aluminium alloy nail at eaves						
	430 × 380 mm tiles	0.09	—	—	—	0.006	ln. m
	418 × 330 mm tiles	0.10	—	—	—	0.007	ln. m
	381 × 227 mm tiles	0.12	—	—	—	0.008	ln. m
H60.055	Verges; 6 mm asbestos cement undercloak strip 150 mm wide; bedded and pointed in cement mortar						
	75 mm lap						
	430 × 380 mm tiles	0.56	—	1.025	0.01	—	ln. m
	418 × 330 mm tiles	0.56	—	1.025	0.01	—	ln. m
	100 mm lap						
	430 × 380 mm tiles	0.58	—	1.025	0.01	—	ln. m
	418 × 330 mm tiles	0.58	—	1.025	0.01	—	ln. m
H60.060	Verges; treble roll left-hand verge tile; 6 mm asbestos cement undercloak strip 150 mm wide; bedded and pointed in cement mortar						
	75 mm lap						
	430 × 380 mm tiles	0.58	3.08	1.025	0.01	0.007	ln. m
	418 × 330 mm tiles	0.58	3.08	1.025	0.01	0.007	ln. m
	100 mm lap						
	430 × 380 mm tiles	0.60	3.30	1.025	0.01	0.007	ln. m
	418 × 330 mm tiles	0.60	3.30	1.025	0.01	0.007	ln. m

		Labour (hr)	Tiles (no.)	Asbestos undercloak (ln. m)	Cement mortar (cu. m)	Aluminium nails (kg)	Unit
H60.065	Ridge tiles; half round; in 457 mm effective lengths; butt jointed; bedded and pointed in cement mortar	1.00	2.30	—	0.03	—	ln. m
H60.070	ends filled with mortar	0.80	—	—	—	—	no.
H60.075	mitred angles	0.24	—	—	—	—	no.
H60.080	mitred three-way intersection	0.80	—	—	—	—	no.

		Labour (hr)	Tiles (no.)	Battens (ln. m)	Roundwire nails (kg)	Roofing felt sq. m	Unit
H60.105	**Roofing; clayware tiles; Sandtoft Goxhill; BS.747 type IF reinforced bitumen felt underlay; 150 mm laps; secured with softwood battens, impregnated, fixed with 65 mm × 3.35 mm round wire nails**						
H60.110	Roof coverings; members at 400 mm centres; 75 mm headlap; 19 × 38 mm battens						
	Old English Pantiles; 342 × 241 mm	0.66	18.96	3.95	0.048	1.15	sq. m
	Gaelic French Tiles; 342 × 241 mm	0.67	19.27	3.95	0.048	1.15	sq. m
	Bold Roll Tiles; 342 × 266 mm	0.64	18.25	3.95	0.048	1.15	sq. m
H60.115	Roof coverings; members at 450 mm centres; 75 mm headlap; 19 × 38 mm battens						
	Old English Pantiles; 342 × 241 mm	0.66	18.96	3.95	0.042	1.15	sq. m
	Gaelic French Tiles; 342 × 241 mm	0.67	19.27	3.95	0.042	1.15	sq. m
	Bold Roll Tiles; 342 × 266 mm	0.64	18.25	3.95	0.042	1.15	sq. m
H60.120	Roof coverings; members at 600 mm centres; 75 mm headlap; 25 × 38 mm battens						
	Old English Pantiles; 342 × 241 mm	0.66	18.96	3.95	0.032	1.15	sq. m
	Gaelic French Tiles; 342 × 241 mm	0.67	19.27	3.95	0.032	1.15	sq. m
	Bold Roll Tiles; 342 × 266 mm	0.64	18.25	3.95	0.032	1.15	sq. m
H60.125	Square cutting at abutment of large openings						
	Old English Pantiles; 342 × 241 mm	0.20	1.90	—	—	—	ln. m
	Gaelic French Tiles; 342 × 241 mm	0.20	1.93	—	—	—	ln. m
	Bold Roll Tiles; 342 × 266 mm	0.20	1.83	—	—	—	ln. m
H60.130	Square cutting at top edges of large openings						
	Old English Pantiles; 342 × 241 mm	0.32	1.90	—	—	—	ln. m
	Gaelic French Tiles; 342 × 241 mm	0.32	1.93	—	—	—	ln. m
	Bold Roll Tiles; 342 × 266 mm	0.32	1.83	—	—	—	ln. m
H60.135	Raking cutting						
	Old English Pantiles; 342 × 241 mm	0.38	2.84	—	—	—	ln. m
	Gaelic French Tiles; 342 × 241 mm	0.38	2.89	—	—	—	ln. m
	Bold Roll Tiles; 342 × 266 mm	0.38	2.74	—	—	—	ln. m
H60.140	Curved cutting						
	Old English Pantiles; 342 × 241 mm	0.46	2.84	—	—	—	ln. m
	Gaelic French Tiles; 342 × 241 mm	0.46	2.89	—	—	—	ln. m
	Bold Roll Tiles; 342 × 266 mm	0.46	2.74	—	—	—	ln. m

		Labour (hr)	Tiles (no.)	Plain tiles (no.)	Cement mortar (cu. m)	Aluminium nails (kg)	Unit
H60.150	Cutting to valleys; bedding and pointing tiles in cement mortar on plain tile slips						
	Old English Pantiles; 342 × 241 mm	0.60	—	6.21	0.01	—	ln. m
	Gaelic French tiles; 342 × 241 mm	0.60	—	6.21	0.01	—	ln. m
	Bold Roll tiles; 342 × 266 mm	0.60	—	6.21	0.01	—	ln. m
H60.205	**Extra over pitched roof coverings for**						
H60.210	Fixing each tile in every alternative course and all perimeter tiles with one 50 × 3.35 mm aluminium alloy nail						
	Old English Pantiles; 342 × 241 mm	0.03	—	—	—	0.013	sq. m
	Gaelic French Tiles; 342 × 241 mm	0.03	—	—	—	0.013	sq. m
	Bold Roll Tiles; 342 × 266 mm	0.03	—	—	—	0.013	sq. m
H60.215	Fixing every tile and all perimeter tiles with one 50 × 3.0 mm aluminium alloy nail						
	Old English Pantiles; 342 × 241 mm	0.06	—	—	—	0.026	sq. m
	Gaelic French Tiles; 342 × 241 mm	0.06	—	—	—	0.026	sq. m
	Bold Roll Tiles; 342 × 266 mm	0.06	—	—	—	0.026	sq. m
H60.220	Extra undercloak plain tiles 265 mm wide at eaves; bedding and pointing in cement mortar						
	Old English Pantiles; 342 × 241 mm	0.32	—	6.21	0.01	—	ln. m
	Gaelic French Tiles; 342 × 241 mm	0.32	—	6.21	0.01	—	ln. m
	Bold Roll Tiles; 342 × 266 mm	0.32	—	6.21	0.01	—	ln. m
H60.225	Verges; single extra undercloak course plain tiles 265 mm wide; bedding and pointing in cement mortar						
	Old English Pantiles; 342 × 241 mm	0.50	—	6.21	0.01	—	ln. m
	Gaelic French Tiles; 342 × 241 mm	0.50	—	6.21	0.01	—	ln. m
	Bold Roll Tiles; 342 × 266 mm	0.50	—	6.21	0.01	—	ln. m
H60.230	Verges; double roll tiles; single extra undercloak course plain tiles 265 mm wide; bedding and pointing in cement mortar						
	Old English Pantiles; 342 × 241 mm	0.50	3.84	6.21	0.01	—	ln. m
	Gaelic French Tiles; 342 × 241 mm	0.50	3.84	6.21	0.01	—	ln. m
	Bold Roll Tiles; 342 × 266 mm	0.50	3.84	6.21	0.01	—	ln. m

	Labour (hr)	Tiles (no.)	Plain tiles (no.)	Cement mortar (cu. m)	Dentil slips (no.)	Unit
H60.235 Ridge tiles, half round; 305 mm effective lengths; double dentil slips both sides; bedding and pointing in cement mortar						
Old English Pantiles	1.24	3.36	—	0.03	10.10	ln. m
Gaelic French Tiles	1.00	3.36	—	0.03	—	ln. m
Bold Roll Tiles	1.24	3.36	—	0.03	10.10	ln. m
H60.240 Extra over ridge tiles for						
ends filled with mortar	0.80	—	—	0.01	—	no.
mitred angles	0.20	0.03	—	—	—	no.
mitred three-way intersection	0.80	0.03	—	—	—	no.

J WATERPROOFING

JA   Notes

I   The labour constants in this section are based on one labourer.

J40 *FLEXIBLE SHEET TANKING/DAMP-PROOF MEMBRANES*

J40.005	**Polythene damp-proof membrane**
J40.010	1200 gauge polythene damp-proof membrane

J40.015	**Bitu-thene waterproofing membrane**
J40.020	1000 gauge Bitu-thene waterproofing membrane

J40   FLEXIBLE SHEET TANKING/DAMP-PROOF MEMBRANES

	Labour (hr)	Polythene sheet (sq. m)	Sealing tape (ln. m)	Bitu-thene (sq. m)	Primer (litre)	Unit
J40.005 **Polythene damp-proof membrane**						
J40.010 1200 gauge polythene damp-proof membrane with 150 mm side and end laps; laid						
horizontal						
over 300 mm wide	0.03	1.100	—	—	—	sq. m
vertical						
not exceeding 150 mm wide	0.01	0.165	—	—	—	ln. m
150–300 mm wide	0.02	0.330	—	—	—	ln. m
Extra for						
welded joints	0.03	—	—	—	—	sq. m
taped joints	0.01	—	0.03	—	—	sq. m
J40.015 **Bitu-thene waterproofing membrane**						
J40.020 1000 gauge Bitu-thene waterproofing membrane with 75 mm side and end laps; priming surfaces with one coat Bitu-thene primer						
horizontal						
over 300 mm wide	0.27	—	—	1.150	0.550	sq. m
vertical						
not exceeding 150 mm wide	0.06	—	—	0.173	0.083	ln. m
150–300 mm wide	0.11	—	—	0.345	0.165	ln. m
over 300 mm wide	0.25	—	—	1.150	0.550	sq. m
internal or external angles						
300 mm wide reinforcing strip	0.07	—	—	0.345	0.165	ln. m

K LININGS/SHEATHING

KA   **Notes**

1   The labour constants in Section K10 are based on two craftsmen plasterers and one labourer (two and
           one plastering gang).

2   The labour constants in Sections K11 and K20 are based on one craftsman joiner.

K10   *PLASTERBOARD DRY LINING*

K10.005	**9.5 mm thick gypsum wallboard**
K10.010	Walls
K10.015	Beams, faces
K10.020	Columns, faces
K10.025	Reveals and soffits of openings and recesses
K10.030	Ceilings

K10.035	**12.7 mm thick gypsum wallboard**
K10.040	Walls
K10.045	Beams, faces
K10.050	Columns, faces
K10.055	Ceilings

K10.105	**9.5 mm thick gypsum wallboard, tapered edges**
K10.110	Walls
K10.115	Beams, faces
K10.120	Columns, faces
K10.125	Reveals and soffits of openings and recesses
K10.130	Ceilings

K10.135	**12.7 mm thick gypsum wallboard, tapered edges**
K10.140	Walls
K10.145	Beams, faces
K10.150	Columns, faces
K10.155	Reveals and soffits of openings and recesses
K10.160	Ceilings
K10.205	**25 mm thick Gyproc thermal board, tapered edges**
K10.210	Walls
K10.215	Reveals and soffits of openings and recesses
K10.220	Ceilings
K10.225	**32 mm thick Gyproc thermal board, tapered edges**
K10.230	Walls
K10.235	Reveals and soffits of openings and recesses
K10.240	Ceilings
K10.305	**40 mm thick Gyproc thermal board, tapered edges**
K10.310	Walls
K10.315	Reveals and soffits of openings and recesses
K10.320	Ceilings

K10.325	**50 mm thick Gyproc thermal board, tapered edges**
K10.330	Walls
K10.335	Reveals and soffits of openings and recesses
K10.340	Ceilings
K10.405	**Gyproc Dry wall top coat, one coat work**
K10.410	Walls
K10.415	Beams, faces
K10.420	Columns, faces
K10.425	Reveals and soffits of openings and recesses
K10.430	Ceilings
K10.505	**Angles to linings**
K10.510	Internal angles
K10.515	External angles
K10.520	**Beads**
K10.525	Angle beads; galvanized steel
K10.530	Stop beads; galvanized steel

K10   PLASTERBOARD DRY LINING

		Labour (hr)	Plaster-board (sq. m)	Galvanized clout nails 30×2.65 (kg)	40×2.65 (kg)	Scrim cloth (ln. m)	Unit
K10.005	**9.5 mm thick gypsum wallboard, BS.1230; 1.5 mm joints filled with plaster and scrimmed; fixing with galvanized nails; on timber base; internal**						
K10.010	Walls						
	1500 mm high	0.30	1.575	1.592	—	2.475	ln. m
	1800 mm high	0.36	1.890	1.910	—	2.970	ln. m
	2100 mm high	0.42	2.205	2.228	—	3.465	ln. m
	2400 mm high	0.48	2.520	2.546	—	3.960	ln. m
	2700 mm high	0.54	2.835	2.865	—	4.455	ln. m
	3000 mm high	0.60	3.150	3.183	—	4.950	ln. m
K10.015	Beams, faces						
	not exceeding 600 mm girth	0.20	0.630	0.037	—	1.000	ln. m
	600–1200 mm girth	0.40	1.260	0.091	—	1.980	ln. m
	1200–1800 mm girth	0.59	1.890	0.137	—	2.970	ln. m
	1800–2400 mm girth	0.79	2.520	0.182	—	3.960	ln. m
K10.020	Columns, faces						
	not exceeding 600 mm girth	0.15	0.630	0.037	—	1.000	ln. m
	600–1200 mm girth	0.30	1.260	0.091	—	1.980	ln. m
	1200–1800 mm girth	0.45	1.890	0.137	—	2.970	ln. m
	1800–2400 mm girth	0.60	2.520	0.182	—	3.960	ln. m
K10.025	Reveals and soffits of openings and recesses						
	not exceeding 300 mm wide	0.20	0.315	0.023	—	0.495	ln. m
	300–600 mm wide	0.20	0.630	0.037	—	1.000	ln. m
K10.030	Ceilings						
	over 300 mm wide	0.29	1.050	0.061	—	1.65	sq. m

			Galvanized clout nails			
		Plaster-board	30x2.65	40x2.65	Scrim cloth	
	Labour (hr)	(sq. m)	(kg)	(kg)	(ln. m)	Unit

K10.035	**12.7 mm thick gypsum wallboard, BS.1230; 1.5 mm joints filled with plaster and scrimmed; fixing with galvanized nails; on timber base; internal**					
K10.040	Walls					
	1500 mm high	0.38	1.575	—	0.095	2.475	ln. m
	1800 mm high	0.45	1.890	—	0.113	2.970	ln. m
	2100 mm high	0.53	2.205	—	0.132	3.465	ln. m
	2400 mm high	0.60	2.520	—	0.151	3.960	ln. m
	2700 mm high	0.68	2.835	—	0.170	4.455	ln. m
	3000 mm high	0.75	3.150	—	0.189	4.950	ln. m
K10.045	Beams, faces					
	not exceeding 600 mm girth	0.25	0.630	—	0.038	1.000	ln. m
	600–1200 mm girth	0.50	1.260	—	0.076	1.980	ln. m
	1200–1800 mm girth	0.76	1.890	—	0.113	2.970	ln. m
	1800–2400 mm girth	1.01	2.520	—	0.151	3.960	ln. m
K10.050	Columns, faces					
	not exceeding 600 mm girth	0.19	0.630	—	0.038	1.000	ln. m
	600–1200 mm girth	0.37	1.260	—	0.076	1.980	ln. m
	1200–1800 mm girth	0.56	1.890	—	0.113	2.970	ln. m
	1800–2400 mm girth	0.74	2.520	—	0.151	3.960	ln. m
K10.055	Reveals and soffits of openings and recesses					
	not exceeding 300 mm wide	0.25	0.315	—	0.024	0.495	ln. m
	300–600 mm wide	0.25	0.630	—	0.038	1.000	ln. m
K10.060	Ceilings					
	over 300 mm wide	0.36	1.050	—	0.063	1.65	sq. m

	Labour (hr)	Plasterboard (sq. m)	Galvanized nails (kg)	Joint filler (kg)	Joint tape (ln. m)	Unit
K10.105	**9.5 mm thick gypsum wallboard, tapered edges; BS.1230; 1.5 mm joints filled with Gyproc joint filler and taped flush; fixing with galvanized nails; nail heads filled with Gyproc joint filler; on timber base; internal**					
K10.110	Walls					
	1500 mm high	0.45	1.575	0.092	0.363	2.261	ln. m
	1800 mm high	0.54	1.890	0.110	0.436	2.713	ln. m
	2100 mm high	0.63	2.205	0.128	0.508	3.165	ln. m
	2400 mm high	0.72	2.520	0.146	0.581	3.617	ln. m
	2700 mm high	0.81	2.835	0.165	0.653	4.069	ln. m
	3000 mm high	0.90	3.150	0.183	0.726	4.521	ln. m
K10.115	Beams, faces					
	not exceeding 600 mm girth	0.30	0.630	0.037	0.145	0.904	ln. m
	600–1200 mm girth	0.60	1.260	0.073	0.290	1.808	ln. m
	1200–1800 mm girth	0.90	1.890	0.110	0.436	2.713	ln. m
	1800–2400 mm girth	1.20	2.520	0.146	0.581	3.617	ln. m
K10.120	Columns, faces					
	not exceeding 600 mm girth	0.28	0.630	0.037	0.145	0.904	ln. m
	600–1200 mm girth	0.56	1.260	0.073	0.290	1.808	ln. m
	1200–1800 mm girth	0.85	1.890	0.110	0.436	2.713	ln. m
	1800–2400 mm girth	1.13	2.520	0.146	0.581	3.617	ln. m
K10.125	Reveals and soffits of openings and recesses					
	not exceeding 300 mm wide	0.30	0.315	0.023	0.076	0.473	ln. m
	300–600 mm wide	0.30	0.630	0.037	0.145	0.904	ln. m
K10.130	Ceilings					
	over 300 mm wide	0.44	1.05	0.061	0.242	1.507	sq. m

	Labour (hr)	Plasterboard (sq. m)	Galvanized nails (kg)	Joint filler (kg)	Joint tape (ln. m)	Unit
K10.135 **12.7 mm thick gypsum wallboard, tapered edges; BS.1230; 1.5 mm joints filled with Gyproc joint filler and taped flush; fixing with galvanized nails; nail heads filled with Gyproc joint filler; on timber base; internal**						
K10.140 Walls						
1500 mm high	0.57	1.575	0.095	0.363	2.261	ln. m
1800 mm high	0.68	1.890	0.113	0.436	2.713	ln. m
2100 mm high	0.80	2.205	0.132	0.508	3.165	ln. m
2400 mm high	0.91	2.520	0.151	0.581	3.617	ln. m
2700 mm high	1.03	2.835	0.170	0.653	4.069	ln. m
3000 mm high	1.14	3.150	0.189	0.726	4.521	ln. m
K10.145 Beams, faces						
not exceeding 600 mm girth	0.38	0.630	0.038	0.145	0.904	ln. m
600–1200 mm girth	0.76	1.260	0.076	0.290	1.808	ln. m
1200–1800 mm girth	1.13	1.890	0.113	0.436	2.713	ln. m
1800–2400 mm girth	1.51	2.520	0.151	0.581	3.617	ln. m
K10.150 Columns, faces						
not exceeding 600 mm girth	0.28	0.630	0.038	0.145	0.904	ln. m
600–1200 mm girth	0.56	1.260	0.076	0.290	1.808	ln. m
1200–1800 mm girth	0.85	1.890	0.113	0.436	2.713	ln. m
1800–2400 mm girth	1.13	2.520	0.151	0.581	3.617	ln. m
K10.155 Reveals and soffits of openings and recesses						
not exceeding 300 mm wide	0.38	0.315	0.023	0.076	0.473	ln. m
300–600 mm wide	0.38	0.630	0.037	0.145	0.904	ln. m
K10.160 Ceilings						
over 300 mm wide	0.54	1.05	0.063	0.242	1.507	sq. m

		Labour (hr)	Thermal board sq. m	Galvanized nails (kg)	Joint filler (kg)	Joint tape (ln. m)	Unit
K10.205	**25 mm Gyproc thermal board, tapered edges; BS.1230; 3 mm joints, filled with Gyproc joint filler and taped flush; fixing with galvanized nails; nail heads filled with Gyproc joint filler; on timber base; internal**						
K10.210	Walls						
	1500 mm high	0.57	1.575	0.090	0.363	2.261	ln. m
	1800 mm high	0.68	1.890	0.108	0.436	2.713	ln. m
	2100 mm high	0.80	2.205	0.126	0.508	3.165	ln. m
	2400 mm high	0.91	2.520	0.144	0.581	3.617	ln. m
	2700 mm high	1.03	2.835	0.162	0.653	4.069	ln. m
	3000 mm high	1.14	3.150	0.180	0.726	4.521	ln. m
K10.215	Reveals and soffits of openings and recesses						
	not exceeding 300 mm wide	0.23	0.315	0.023	0.076	0.473	ln. m
	300–600 mm wide	0.23	0.630	0.037	0.145	0.904	ln. m
K10.220	Ceilings						
	over 300 mm wide	0.55	1.05	0.060	0.242	1.507	sq. m
K10.225	**32 mm Gyproc thermal board, tapered edges; BS.1230; 3 mm joints, filled with Gyproc joint filler and taped flush; fixing with galvanized nails; nail heads filled with Gyproc joint filler; on timber base; internal**						
K10.230	Walls						
	1500 mm high	0.57	1.575	0.144	0.363	2.261	ln. m
	1800 mm high	0.68	1.890	0.173	0.436	2.713	ln. m
	2100 mm high	0.80	2.205	0.202	0.508	3.165	ln. m
	2400 mm high	0.91	2.520	0.230	0.581	3.617	ln. m
	2700 mm high	1.03	2.835	0.259	0.653	4.069	ln. m
	3000 mm high	1.14	3.150	0.288	0.726	4.521	ln. m
K10.235	Reveals and soffits of openings and recesses						
	not exceeding 300 mm wide	0.23	0.315	0.036	0.076	0.473	ln. m
	300–600 mm wide	0.23	0.630	0.058	0.145	0.904	ln. m
K10.240	Ceilings						
	over 300 mm wide	0.55	1.05	0.096	0.242	1.507	sq. m

		Labour (hr)	Thermal board (sq. m)	Galvanized nails (kg)	Joint filler (kg)	Joint tape (ln. m)	Unit
K10.305	**40 mm Gyproc thermal board, tapered edges; BS.1230; 3 mm joints, filled with Gyproc joint filler and taped flush; fixing with galvanized nails; nail heads filled with Gyproc joint filler; on timber base; internal**						
K10.310	Walls						
	1500 mm high	0.66	1.575	0.144	0.363	2.261	ln. m
	1800 mm high	0.79	1.890	0.173	0.436	2.713	ln. m
	2100 mm high	0.92	2.205	0.202	0.508	3.165	ln. m
	2400 mm high	1.06	2.520	0.230	0.581	3.617	ln. m
	2700 mm high	1.19	2.835	0.259	0.653	4.069	ln. m
	3000 mm high	1.32	3.150	0.288	0.726	4.521	ln. m
K10.315	Reveals and soffits of openings and recesses						
	not exceeding 300 mm wide	0.26	0.315	0.036	0.076	0.473	ln. m
	300–600 mm wide	0.26	0.630	0.058	0.145	0.904	ln. m
K10.320	Ceilings						
	over 300 mm wide	0.64	1.05	0.096	0.242	1.507	sq. m
K10.325	**50 mm Gyproc thermal board, tapered edges; BS.1230; 3 mm joints, filled with Gyproc joint filler and taped flush; fixing with galvanized nails; nail heads filled with Gyproc joint filler; on timber base; internal**						
K10.330	Walls						
	1500 mm high	0.66	1.575	0.200	0.363	2.261	ln. m
	1800 mm high	0.79	1.890	0.239	0.436	2.713	ln. m
	2100 mm high	0.92	2.205	0.279	0.508	3.165	ln. m
	2400 mm high	1.06	2.520	0.319	0.581	3.617	ln. m
	2700 mm high	1.19	2.835	0.359	0.653	4.069	ln. m
	3000 mm high	1.32	3.150	0.399	0.726	4.521	ln. m
K10.335	Reveals and soffits of openings and recesses						
	not exceeding 300 mm wide	0.26	0.315	0.050	0.076	0.473	ln. m
	300–600 mm wide	0.26	0.630	0.080	0.145	0.904	ln. m
K10.340	Ceilings						
	over 300 mm wide	0.64	1.05	0.133	0.242	1.507	sq. m

	Labour (hr)			Joint finish (kg)	Dry wall top coat (litre)	Unit
K10.405	**Gyproc dry wall top coat, one coat work**					
K10.410	Walls on plasterboard base					
	1500 mm high	0.21	—	—	0.545	0.147	ln. m
	1800 mm high	0.25	—	—	0.653	0.176	ln. m
	2100 mm high	0.29	—	—	0.762	0.206	ln. m
	2400 mm high	0.34	—	—	0.871	0.235	ln. m
	2700 mm high	0.38	—	—	0.980	0.265	ln. m
	3000 mm high	0.42	—	—	1.089	0.294	ln. m
K10.415	Beams, faces					
	not exceeding 600 mm girth	0.08	—	—	0.218	0.059	ln. m
	600–1200 mm girth	0.17	—	—	0.436	0.118	ln. m
	1200–1800 mm girth	0.25	—	—	0.653	0.176	ln. m
	1800–2400 mm girth	0.34	—	—	0.871	0.235	ln. m
K10.420	Columns, faces					
	not exceeding 600 mm girth	0.08	—	—	0.218	0.059	ln. m
	600–1200 mm girth	0.17	—	—	0.436	0.118	ln. m
	1200–1800 mm girth	0.25	—	—	0.653	0.176	ln. m
	1800–2400 mm girth	0.34	—	—	0.871	0.235	ln. m
K10.425	Reveals and soffits of openings and recesses					
	not exceeding 300 mm wide	0.08	—	—	0.114	0.029	ln. m
	300–600 mm wide	0.08	—	—	0.228	0.058	ln. m
K10.430	Ceilings					
	over 300 mm wide	0.14	—	—	0.363	0.098	sq. m

		Labour (hr)		Joint finish (kg)	Dry wall top coat (litre)	Beads (ln. m)	Unit
K10.505	**Angles to linings**						
K10.510	Internal angles						
	9.5 mm wallboard	0.14	—	—	1.10	—	ln. m
	12.7 mm wallboard	0.14	—	—	1.10	—	ln. m
K10.515	External angles						
	9.5 mm wallboard	0.17	—	0.006	1.10	—	ln. m
	12.7 mm wallboard	0.17	—	0.006	1.10	—	ln. m
K10.520	**Beads**						
K10.525	Angle beads; galvanized steel; fixing with nails to timber base; Expamet						
	Ref: 553, 3 mm deep	0.13	—	—	—	1.05	ln. m
	Ref: 554, 6 mm deep	0.13	—	—	—	1.05	ln. m
K10.530	Stop beads; galvanized steel; fixing with nails to timber base; Expamet						
	Ref: 560, 3 mm deep	0.10	—	—	—	1.05	ln. m
	Ref: 561, 6 mm deep	0.10	—	—	—	1.05	ln. m

K11   *RIGID SHEET FLOORING/LININGS/CASINGS*

K11.005   **Plywood; BS.1455, Grade 2/3 WBP bonded**

K11.010   Boarding to floors or roofs; butt joints

K11.015   Linings or casings

K11.020   Labour on plywood boarding

K11.105   **Chipboard to BS.5669; flooring grade**

K11.110   Boarding to floors; butt joints

K11.115   Boarding to floors; tongued and grooved joints

K11.120   Boarding to roofs; butt joints

K11.125   Linings or casings

K11.130   Labour on chipboard

K11.205   **Blockboard, BS.3444, 2/2 Grade, BR bonded**

K11.210   Linings or casings

K11.215   Labour on blockboard

K11.305   **Non-asbestos board, BS.476; 'Supalux'**

K11.310   Linings or casings

K11.315   Labour on 'Supalux' boarding

K11.405   **Hardboard, BS.1142; butt joints**

K11.410   Linings or casings

K11.415   Labour on hardboard

K11.505   **Melamine faced chipboard; butt joints**

K11.510   Linings or casings

K11.515   Labour on melamine faced chipboard

K11.605   **Insulation board, BS.1142; butt joints**

K11.610   Linings or casings

K11.615   Labour on insulation board

K11.705   **Plastic laminate, fixing with adhesive to plywood or blockboard base**

K11.710   Linings or casings

K11.715   Edging

K11.720   Wrought hardwood lipping pinned and glued to edge of boarding

K11 RIGID SHEET FLOORING/LININGS/CASINGS

	Labour (hr)			Boarding (sq. m)	Nails (kg)	Unit
K11.005	**Plywood; BS.1455, Grade 2/3 WBP bonded**					
K11.010	Boarding to floors or roofs; butt joints					
12 mm thick	0.36	—	—	1.15	0.06	sq. m
18 mm thick	0.38	—	—	1.15	0.06	sq. m
25 mm thick	0.43	—	—	1.15	0.06	sq. m
K11.015	Linings or casings					
4 mm thick						
not exceeding 100 mm wide	0.14	—	—	0.12	0.01	ln. m
100–200 mm wide	0.16	—	—	0.23	0.01	ln. m
200–300 mm wide	0.18	—	—	0.35	0.02	ln. m
over 300 mm wide	0.35	—	—	1.15	0.06	sq. m
6 mm thick						
not exceeding 100 mm wide	0.14	—	—	0.12	0.01	ln. m
100–200 mm wide	0.16	—	—	0.23	0.01	ln. m
200–300 mm wide	0.18	—	—	0.35	0.02	ln. m
over 300 mm wide	0.35	—	—	1.15	0.06	sq. m
9 mm thick						
not exceeding 100 mm wide	0.17	—	—	0.12	0.01	ln. m
100–200 mm wide	0.18	—	—	0.23	0.01	ln. m
200–300 mm wide	0.21	—	—	0.35	0.02	ln. m
over 300 mm wide	0.41	—	—	1.15	0.06	sq. m
12 mm thick						
not exceeding 100 mm wide	0.20	—	—	0.12	0.01	ln. m
100–200 mm wide	0.22	—	—	0.23	0.01	ln. m
200–300 mm wide	0.25	—	—	0.35	0.02	ln. m
over 300 mm wide	0.50	—	—	1.15	0.06	sq. m
K11.020	Labour on plywood boarding					
4 mm thick						
raking cutting	0.25	—	—	—	—	ln. m
curved cutting	0.50	—	—	—	—	ln. m
notches per 25 mm girth	0.17	—	—	—	—	no.
forming openings, not exceeding 0.50 m^2	0.34	—	—	—	—	no.
6 mm thick						
raking cutting	0.25	—	—	—	—	ln. m
curved cutting	0.50	—	—	—	—	ln. m
notches, per 25 mm girth	0.17	—	—	—	—	no.
forming openings, not exceeding 0.50 m^2	0.34	—	—	—	—	no.

		Labour (hr)			Boarding (sq. m)	Nails (kg)	Unit
K11.020	(continued)						
	9 mm thick						
	raking cutting	0.29	—	—	—	—	ln. m
	curved cutting	0.58	—	—	—	—	ln. m
	notches, per 25 mm girth	0.19	—	—	—	—	no.
	forming openings, not exceeding 0.50 m^2	0.38	—	—	—	—	no.
	12 mm thick						
	raking cutting	0.32	—	—	—	—	ln. m
	curved cutting	0.64	—	—	—	—	ln. m
	notches, per 25 mm girth	0.21	—	—	—	—	no.
	forming openings, not exceeding 0.50 m^2	0.42	—	—	—	—	no.
	18 mm thick						
	raking cutting	0.35	—	—	—	—	ln. m
	curved cutting	0.70	—	—	—	—	ln. m
	notches, per 25 mm girth	0.23	—	—	—	—	no.
	forming openings, not exceeding 0.50 m^2	0.46	—	—	—	—	no.
	25 mm thick						
	raking cutting	0.37	—	—	—	—	ln. m
	curved cutting	0.74	—	—	—	—	ln. m
	notches, per 25 mm girth	0.24	—	—	—	—	no.
	forming openings, not exceeding 0.50 m^2	0.49	—	—	—	—	no.

		Labour (hr)			Boarding (sq. m)	Nails (kg)	Unit
K11.105	**Chipboard to BS 5669; flooring grade**						
K11.110	Boarding to floors; butt joints						
	18 mm thick	0.33	—	—	1.15	0.05	sq. m
	22 mm thick	0.40	—	—	1.15	0.05	sq. m
K11.115	Boarding to floors; tongued and grooved joints						
	18 mm thick	0.40	—	—	1.15	0.05	sq. m
	22 mm thick	0.50	—	—	1.15	0.05	sq. m
K11.120	Boarding to roofs; butt joints						
	18 mm flat to falls	0.33	—	—	1.15	0.05	sq. m
	18 mm sloping	0.40	—	—	1.15	0.05	sq. m
	22 mm flat to falls	0.40	—	—	1.15	0.05	sq. m
	22 mm sloping	0.50	—	—	1.15	0.05	sq. m
K11.125	Linings or casings						
	12 mm thick						
	not exceeding 100 mm wide	0.18	—	—	0.12	0.01	ln. m
	100–200 mm wide	0.21	—	—	0.23	0.01	ln. m
	200–300 mm wide	0.23	—	—	0.35	0.02	ln. m
	over 300 mm wide	0.46	—	—	1.15	0.05	sq. m
	18 mm thick						
	not exceeding 100 mm wide	0.24	—	—	0.12	0.01	ln. m
	100–200 mm wide	0.27	—	—	0.23	0.01	ln. m
	200–300 mm wide	0.30	—	—	0.35	0.02	ln. m
	over 300 mm wide	0.61	—	—	1.15	0.05	sq. m
K11.130	Labour on chipboard						
	12 mm thick						
	raking cutting	0.29	—	—	—	—	ln. m
	curved cutting	0.58	—	—	—	—	ln. m
	notches, per 25 mm girth	0.19	—	—	—	—	no.
	forming openings, not exceeding 0.50 m^2	0.38	—	—	—	—	no.
	18 mm thick						
	raking cutting	0.29	—	—	—	—	ln. m
	curved cutting	0.58	—	—	—	—	ln. m
	notches, per 25 mm girth	0.19	—	—	—	—	no.
	forming openings, not exceeding 0.50 m^2	0.38	—	—	—	—	no.
	22 mm thick						
	raking cutting	0.32	—	—	—	—	ln. m
	curved cutting	0.64	—	—	—	—	ln. m
	notches, per 25 mm girth	0.21	—	—	—	—	no.
	forming openings, not exceeding 0.05 m^2	0.42	—	—	—	—	no.

	Labour (hr)			Boarding (sq. m)	Nails (kg)	Unit
K11.205	**Blockboard, BS.3444, 2/2 grade, BR bonded; butt joints**					
K11.210	Linings or casings					
	12 mm thick					
	not exceeeding 100 mm wide	0.18	—	—	0.12	—	ln. m
	100–200 mm wide	0.20	—	—	0.23	—	ln. m
	200–300 mm wide	0.23	—	—	0.35	—	ln. m
	over 300 mm wide	0.46	—	—	1.15	—	sq. m
	18 mm thick;					
	not exceeding 100 mm wide	0.24	—	—	0.12	—	ln. m
	100–200 mm wide	0.27	—	—	0.23	—	ln. m
	200–300 mm wide	0.30	—	—	0.35	—	ln. m
	over 300 mm wide	0.61	—	—	1.15	—	sq. m
	25 mm thick					
	not exceeding 100 mm wide	0.30	—	—	0.12	—	ln. m
	100–200 mm wide	0.34	—	—	0.23	—	ln. m
	200–300 mm wide	0.38	—	—	0.35	—	ln. m
	over 300 mm wide	0.76	—	—	1.15	—	sq. m
K11.215	Labour on blockboard					
	12 mm thick					
	raking cutting	0.32	—	—	—	—	ln. m
	curved cutting	0.64	—	—	—	—	ln. m
	notches, per 25 mm girth	0.21	—	—	—	—	no.
	forming openings, not exceeding 0.50 m^2	0.42	—	—	—	—	no.
	18 mm thick					
	raking cutting	0.32	—	—	—	—	ln. m
	curved cutting	0.64	—	—	—	—	ln. m
	notches, per 25 mm girth	0.21	—	—	—	—	no.
	forming openings, not exceeding 0.50 m^2	0.42	—	—	—	—	no.
	25 mm thick					
	raking cutting	0.35	—	—	—	—	ln. m
	curved cutting	0.70	—	—	—	—	ln. m
	notches, per 25 mm girth	0.23	—	—	—	—	no.
	forming openings, not exceeding 0.50 m^2	0.46	—	—	—	—	no.

	Labour (hr)			Boarding (sq. m)	Nails (kg)	Unit
K11.305	**Non-asbestos board, BS.476; 'Supalux'; butt joints**					
K11.310	Linings or casings					
	6 mm thick					
	not exceeding 100 mm wide	0.23	—	—	0.12	—	ln. m
	100–200 mm wide	0.25	—	—	0.23	—	ln. m
	200–300 mm wide	0.29	—	—	0.35	—	ln. m
	over 300 mm wide	0.57	—	—	1.15	—	sq. m
	9 mm thick					
	not exceeding 100 mm wide	0.27	—	—	0.12	—	ln. m
	100–200 mm wide	0.30	—	—	0.23	—	ln. m
	200–300 mm wide	0.34	—	—	0.35	—	ln. m
	over 300 mm wide	0.67	—	—	1.15	—	sq. m
	12 mm thick					
	not exceeding 100 mm wide	0.33	—	—	0.12	—	ln. m
	100–200 mm wide	0.36	—	—	0.23	—	ln. m
	200–300 mm wide	0.41	—	—	0.35	—	ln. m
	over 300 mm wide	0.82	—	—	1.15	—	sq. m
K11.315	Labour on 'Supalux' boarding					
	6 mm thick					
	raking cutting	0.29	—	—	—	—	ln. m
	curved cutting	0.58	—	—	—	—	ln. m
	notches, per 25 mm girth	0.19	—	—	—	—	no.
	forming openings, not exceeding 0.50 m^2	0.38	—	—	—	—	no.
	9 mm thick					
	raking cutting	0.29	—	—	—	—	ln. m
	curved cutting	0.58	—	—	—	—	ln. m
	notches, per 25 mm girth	0.19	—	—	—	—	no.
	forming openings, not exceeding 0.50 m^2	0.38	—	—	—	—	no.
	12 mm thick					
	raking cutting	0.29	—	—	—	—	ln. m
	curved cutting	0.58	—	—	—	—	ln. m
	notches, per 25 mm girth	0.19	—	—	—	—	no.
	forming openings, not exceeding 0.50 m^2	0.38	—	—	—	—	no.

		Labour (hr)			Boarding (sq. m)	Nails (kg)	Unit
KII.405	**Hardboard, BS.1142; butt joints**						
KII.410	Linings or casings						
	3.2 mm thick						
	not exceeding 100 mm wide	0.14	—	—	0.12	—	ln. m
	100–200 mm wide	0.16	—	—	0.23	—	ln. m
	200–300 mm wide	0.18	—	—	0.35	—	ln. m
	over 300 mm wide	0.35	—	—	1.15	—	sq. m
	6.4 mm thick						
	not exceeding 100 mm wide	0.14	—	—	0.12	—	ln. m
	100–200 mm wide	0.16	—	—	0.23	—	ln. m
	200–300 mm wide	0.18	—	—	0.35	—	ln. m
	over 300 mm wide	0.35	—	—	1.15	—	sq. m
KII.415	Labour on hardboard						
	3.2 mm thick						
	raking cutting	0.25	—	—	—	—	ln. m
	curved cutting	0.50	—	—	—	—	ln. m
	notches, per 25 mm girth	0.14	—	—	—	—	no.
	forming openings, not exceeding 0.50 m^2	0.36	—	—	—	—	no.
	6.4 mm thick						
	raking cutting	0.25	—	—	—	—	ln. m
	curved cutting	0.50	—	—	—	—	ln. m
	notches, per 25 mm girth	0.14	—	—	—	—	no.
	forming openings, not exceeding 0.50 m^2	0.36	—	—	—	—	no.

	Labour (hr)			Boarding (sq. m)	Nails (kg)	Unit
K11.505	**Melamine faced chipboard; butt joints**					
K11.510	Linings or casings					
	15 mm thick					
	not exceeding 100 mm wide	0.23	—	—	0.12	—	ln. m
	100–200 mm wide	0.26	—	—	0.23	—	ln. m
	200–300 mm wide	0.29	—	—	0.35	—	ln. m
	over 300 mm wide	0.58	—	—	1.15	—	sq. m
	18 mm thick					
	not exceeding 100 mm wide	0.31	—	—	0.12	—	ln. m
	100–200 mm wide	0.34	—	—	0.23	—	ln. m
	200–300 mm wide	0.38	—	—	0.35	—	ln. m
	over 300 mm wide	0.77	—	—	1.15	—	sq. m
K11.515	Labour on melamine faced chipboard					
	15 mm thick					
	raking cutting	0.29	—	—	—	—	ln. m
	curved cutting	0.58	—	—	—	—	ln. m
	notches, per 25 mm girth	0.19	—	—	—	—	no.
	forming openings, not exceeding 0.50 m^2	0.38	—	—	—	—	no.
	18 mm thick					
	raking cutting	0.29	—	—	—	—	ln. m
	curved cutting	0.58	—	—	—	—	ln. m
	notches, per 25 mm girth	0.19	—	—	—	—	no.
	forming openings, not exceeding 0.50 m^2	0.38	—	—	—	—	no.

	Labour (hr)			Boarding (sq. m)	Nails (kg)	Unit
K11.605	**Insulation board, BS.1142; butt joints**					
K11.610	Linings or casings					
12 mm thick						
not exceeding 100 mm wide	0.13	—	—	0.12	—	ln. m
100–200 mm wide	0.15	—	—	0.23	—	ln. m
200–300 mm wide	0.17	—	—	0.35	—	ln. m
over 300 mm wide	0.33	—	—	1.15	—	sq. m
19 mm thick						
not exceeding 100 mm wide	0.14	—	—	0.12	—	ln. m
100–200 mm wide	0.16	—	—	0.23	—	ln. m
200–300 mm wide	0.19	—	—	0.35	—	ln. m
over 300 mm wide	0.35	—	—	1.15	—	sq. m
25 mm thick						
not exceeding 100 mm wide	0.18	—	—	0.12	—	ln. m
100–200 mm wide	0.20	—	—	0.23	—	ln. m
200–300 mm wide	0.22	—	—	0.35	—	ln. m
over 300 mm wide	0.44	—	—	1.15	—	sq. m
K11.615	Labour on insulation board					
12 mm thick						
raking cutting	0.12	—	—	—	—	ln. m
curved cutting	0.24	—	—	—	—	ln. m
notches, per 25 mm girth	0.06	—	—	—	—	no.
forming openings, not exceeding 0.50 m²	0.25	—	—	—	—	no.
19 mm thick						
raking cutting	0.12	—	—	—	—	ln. m
curved cutting	0.24	—	—	—	—	ln. m
notches, per 25 mm girth	0.06	—	—	—	—	no.
forming openings, not exceeding 0.50 m²	0.25	—	—	—	—	no.
25 mm thick						
raking cutting	0.17	—	—	—	—	ln. m
curved cutting	0.34	—	—	—	—	ln. m
notches, per 25 mm girth	0.12	—	—	—	—	no.
forming openings, not exceeding 0.50 m²	0.25	—	—	—	—	no.

	Labour (hr)		Boarding (sq. m)	Adhesive (litre)	Edging (ln. m)	Unit
K11.705	**Plastic laminate, fixing with adhesive to plywood or blockboard base**					
K11.710	Linings or casings; 1.5 mm thick					
	not exceeding 100 mm wide	0.80	—	0.12	0.040	—	ln. m
	100–200 mm wide	0.89	—	0.23	0.070	—	ln. m
	200–300 mm wide	1.00	—	0.35	0.110	—	ln. m
	over 300 mm wide	2.00	—	1.15	0.360	—	sq. m
K11.715	Edging; 1.5 mm thick					
	12 mm wide	0.40	—	—	0.010	1.05	ln. m
	18 mm wide	0.40	—	—	0.010	1.05	ln. m
	25 mm wide	0.40	—	—	0.010	1.05	ln. m
K11.720	Wrought hardwood lipping pinned and glued to edge of boarding; 6 mm thick					
	12 mm wide	0.20	—	—	0.004	1.075	ln. m
	15 mm wide	0.20	—	—	0.005	1.075	ln. m
	18 mm wide	0.20	—	—	0.006	1.075	ln. m
	25 mm wide	0.20	—	—	0.009	1.075	ln. m

K20   *TIMBER BOARD FLOORING/SHEATHING*

K20.005	**Wrought softwood**
K20.010	Boarding to walls; 100 mm width boards; tongued, grooved and V-jointed
K20.015	Boarding to floors; 125 mm width boards; tongued and grooved joints
K20.020	Boarding to ceilings; 100 mm width boards; tongued, grooved and V-jointed
K20.025	Boarding to roofs; 150 mm width boards; butt jointed
K20.030	Boarding to roofs; 125 mm width boards; tongued and grooved joints
K20.035	Boarding to tops and cheeks of dormers; 150 mm width boards; butt joints
K20.040	Boarding to tops and cheeks of dormers; 125 mm width boards; tongued and grooved joints
K20.045	Raking cutting on boarding
K20.050	Curved cutting on boarding
K20.055	Mitred angles on boarding
K20.060	Notches; per 25 mm girth on boarding
K20.065	Cutting and fitting around obstructions; per 25 mm girth on boarding

K20   TIMBER BOARD FLOORING/SHEATHING

	Labour (hr)			Timber (ln. m)	Nails (kg)	Unit
K20.005 **Wrought softwood**						
K20.010 Boarding to walls; 100 mm nominal width boards; tongued, grooved and V-jointed, one side, matchboarding						
13 mm thick	1.18	—	—	12.04	0.40	sq. m
19 mm thick	1.18	—	—	12.04	0.40	sq. m
diagonally						
13 mm thick	1.30	—	—	12.04	0.40	sq. m
19 mm thick	1.30	—	—	12.04	0.40	sq. m
K20.015 Boarding to floors; 125 mm nominal width boards; tongued and grooved joints						
19 mm thick	0.88	—	—	9.43	0.32	sq. m
25 mm thick	0.88	—	—	9.43	0.32	sq. m
K20.020 Boarding to ceilings; 100 mm nominal width boards; tongued, grooved and V-jointed, one side, matchboarding						
13 mm thick	1.59	—	—	12.04	0.40	sq. m
19 mm thick	1.59	—	—	12.04	0.40	sq. m
diagonally						
13 mm thick	1.89	—	—	12.04	0.40	sq. m
19 mm thick	1.89	—	—	12.04	0.40	sq. m
K20.025 Boarding to roofs; 150 mm nominal width boards; butt joints						
flat to falls						
19 mm thick	0.67	—	—	7.17	0.25	sq. m
25 mm thick	0.67	—	—	7.17	0.25	sq. m
flat to falls; diagonally						
19 mm thick	0.80	—	—	7.17	0.25	sq. m
25 mm thick	0.80	—	—	7.17	0.25	sq. m
to slope						
19 mm thick	0.71	—	—	7.17	0.25	sq. m
25 mm thick	0.71	—	—	7.17	0.25	sq. m
to slope; diagonally						
19 mm thick	0.83	—	—	7.17	0.25	sq. m
25 mm thick	0.83	—	—	7.17	0.25	sq. m

	Labour (hr)			Timber (ln. m)	Nails (kg)	Unit
K20.030 Boarding to roofs; 125 mm nominal width boards; tongued and grooved joints						
flat to falls						
19 mm thick	0.88	—	—	9.35	0.32	sq. m
25 mm thick	0.88	—	—	9.35	0.32	sq. m
flat to falls; diagonally						
19 mm thick	1.03	—	—	9.35	0.32	sq. m
25 mm thick	1.03	—	—	9.35	0.32	sq. m
to slope						
19 mm thick	0.93	—	—	9.35	0.32	sq. m
25 mm thick	0.93	—	—	9.35	0.32	sq. m
to slope; diagonally						
19 mm thick	1.09	—	—	9.35	0.32	sq. m
25 mm thick	1.09	—	—	9.35	0.32	sq. m
K20.035 Boarding to tops and cheeks of dormers; 150 mm nominal width boards; butt joints						
flat to falls						
19 mm thick	0.89	—	—	7.17	0.25	sq. m
25 mm thick	0.89	—	—	7.17	0.25	sq. m
flat to falls; diagonally						
19 mm thick	1.05	—	—	7.17	0.25	sq. m
25 mm thick	1.05	—	—	7.17	0.25	sq. m
to slope or vertical						
19 mm thick	0.95	—	—	7.17	0.25	sq. m
25 mm thick	0.95	—	—	7.17	0.25	sq. m
to slope or vertical; diagonally						
19 mm thick	1.11	—	—	7.17	0.25	sq. m
25 mm thick	1.11	—	—	7.17	0.25	sq. m
K20.040 Boarding to tops and cheeks of dormers; 125 mm nominal width boards; tongued and grooved joints						
flat to falls						
19 mm thick	1.16	—	—	9.43	0.32	sq. m
25 mm thick	1.16	—	—	9.43	0.32	sq. m
flat to falls; diagonally						
19 mm thick	1.37	—	—	9.43	0.32	sq. m
25 mm thick	1.37	—	—	9.43	0.32	sq. m
to slope or vertical						
19 mm thick	1.22	—	—	9.43	0.32	sq. m
25 mm thick	1.22	—	—	9.43	0.32	sq. m
to slope or vertical; diagonally						
19 mm thick	1.43	—	—	9.43	0.32	sq. m
25 mm thick	1.43	—	—	9.43	0.32	sq. m

		Labour (hr)			Timber (ln. m)	Nails (kg)	Unit
K20.045	Raking cutting on boarding						
	13 mm thick	0.13	—	—	—	—	ln. m
	19 mm thick	0.13	—	—	—	—	ln. m
	25 mm thick	0.13	—	—	—	—	ln. m
K20.050	Curved cutting on boarding						
	13 mm thick	0.29	—	—	—	—	ln. m
	19 mm thick	0.29	—	—	—	—	ln. m
	25 mm thick	0.29	—	—	—	—	ln. m
K20.055	Mitred angles on boarding						
	13 mm thick	0.57	—	—	—	—	ln. m
	19 mm thick	0.57	—	—	—	—	ln. m
	25 mm thick	0.57	—	—	—	—	ln. m
K20.060	Notches; per 25 mm girth on boarding						
	13 mm thick	0.17	—	—		—	no.
	19 mm thick	0.20	—	—		—	no.
	25 mm thick	0.25	—	—		—	no.
K20.065	Cutting and fitting around obstructions; per 25 mm girth on boarding						
	13 mm thick	0.20	—	—	—	—	no.
	19 mm thick	0.25	—	—	—	—	no.
	25 mm thick	0.30	—	—	—	—	no.

L WINDOWS/DOORS/STAIRS

LA	**Notes**
1	The labour constants in Sections L10, L20, L21 and L30 are based on one craftsman joiner.
2	The labour constants in Section L40 are based on one craftsman joiner, glazier or plumber.

L10 *TIMBER WINDOWS/ROOFLIGHTS*

L10.005	**Standard softwood windows**
L10.010	Metric casement windows
L10.015	Double hung sliding sash windows
L10.020	Georgian type bow windows
L10.105	**Velux roof windows**
L10.110	Velux windows
L10.115	**Velux flashing units to suit windows**
L10.120	Flashing type U, L, VP-U and VP-L
L10.125	Flashing type UBU 100; to suit pair of side-by-side windows

L10   TIMBER WINDOWS/ROOFLIGHTS

		Labour (hr)		Frames (no.)	Nails (kg)	Screws (no.)	Unit
L10.005	**Standard softwood windows**						
L10.010	Metric casement windows						
	600 × 750 mm	0.33	—	1.00	—	—	no.
	600 × 900 mm	0.35	—	1.00	—	—	no.
	600 × 1050 mm	0.37	—	1.00	—	—	no.
	600 × 1200 mm	0.43	—	1.00	—	—	no.
	600 × 1350 mm	0.52	—	1.00	—	—	no.
	1200 × 750 mm	0.75	—	1.00	—	—	no.
	1200 × 900 mm	0.86	—	1.00	—	—	no.
	1200 × 1050 mm	0.99	—	1.00	—	—	no.
	1200 × 1200 mm	1.14	—	1.00	—	—	no.
	1200 × 1350 mm	1.31	—	1.00	—	—	no.
	1800 × 750 mm	1.08	—	1.00	—	—	no.
	1800 × 900 mm	1.30	—	1.00	—	—	no.
	1800 × 1050 mm	1.50	—	1.00	—	—	no.
	1800 × 1200 mm	1.73	—	1.00	—	—	no.
	1800 × 1350 mm	1.99	—	1.00	—	—	no.
	2400 × 1050 mm	2.17	—	1.00	—	—	no.
	2400 × 1200 mm	2.50	—	1.00	—	—	no.
	2400 × 1350 mm	2.75	—	1.00	—	—	no.
L10.015	Double hung sliding sash windows						
	597 × 1102 mm	0.67	—	1.00	—	—	no.
	597 × 1406 mm	0.77	—	1.00	—	—	no.
	597 × 1711 mm	0.89	—	1.00	—	—	no.
	825 × 1102 mm	1.00	—	1.00	—	—	no.
	825 × 1406 mm	1.50	—	1.00	—	—	no.
	825 × 1711 mm	1.65	—	1.00	—	—	no.
	1054 × 1102 mm	1.25	—	1.00	—	—	no.
	1054 × 1406 mm	1.75	—	1.00	—	—	no.
	1054 × 1711 mm	1.93	—	1.00	—	—	no.
L10.020	Georgian type bow windows						
	1524 × 1378 × 137 mm projection	1.78	—	1.00	—	—	no.
	1829 × 1378 × 137 mm projection	2.01	—	1.00	—	—	no.
	2438 × 1378 × 273 mm projection	2.76	—	1.00	—	—	no.
	2978 × 1378 × 413 mm projection	3.45	—	1.00	—	—	no.

		Labour (hr)	Frames (no.)	Flashing (no.)	Screws (no.)	Unit
L10.105	**Velux roof windows in laminated pine construction, impregnated with clear preservative; anodized aluminium external facing; factory glazed with sealed double glazed units; screwing frames to softwood roof timbers**					
L10.110	Velux windows, type					
	GGL-1, 780 × 980 mm	1.17	—	1.00	—	—	no.
	GGL-2, 780 × 1400 mm	1.40	—	1.00	—	—	no.
	GGL-3, 940 × 1600 mm	1.75	—	1.00	—	—	no.
	GGL-4, 1140 × 1180 mm	1.50	—	1.00	—	—	no.
	GGL-5, 700 × 1180 mm	1.17	—	1.00	—	—	no.
	GGL-6, 550 × 980 mm	0.92	—	1.00	—	—	no.
	GGL-7, 1340 × 980 mm	1.58	—	1.00	—	—	no.
	GGL-8, 1340 × 1400 mm	2.33	—	1.00	—	—	no.
	GGL-9, 550 × 700 mm	0.67	—	1.00	—	—	no.
L10.115	**Velux flashing units to suit windows**					
L10.120	Flashing type U, L, VP-U and VP-L; to suit windows					
	GGL-1	1.17	—	—	1.00	—	no.
	GGL-2	1.40	—	—	1.00	—	no.
	GGL-3	1.75	—	—	1.00	—	no.
	GGL-4	1.50	—	—	1.00	—	no.
	GGL-5	1.17	—	—	1.00	—	no.
	GGL-6	0.92	—	—	1.00	—	no.
	GGL-7	1.58	—	—	1.00	—	no.
	GGL-8	2.33	—	—	1.00	—	no.
	GGL-9	0.67	—	—	1.00	—	no.
L10.125	Flashing type UBU 100; to suit pair of side-by-side windows; type					
	GGL-1	2.08	—	—	1.00	—	no.
	GGL-2	2.29	—	—	1.00	—	no.
	GGL-3	3.21	—	—	1.00	—	no.
	GGL-4	2.75	—	—	1.00	—	no.
	GGL-5	2.08	—	—	1.00	—	no.
	GGL-6	1.58	—	—	1.00	—	no.
	GGL-7	2.92	—	—	1.00	—	no.
	GGL-8	4.23	—	—	1.00	—	no.
	GGL-9	1.17	—	—	1.00	—	no.

L20   *TIMBER DOORS*

L20.005	**External doors**
L20.010	Panelled doors; hardwood
L20.015	Panelled doors; softwood
L20.020	Casement doors (pairs)
L20.025	Framed, ledged and braced doors
L20.030	Ledged and braced doors
L20.035	Stable doors
L20.040	Plywood-faced flush doors
L20.045	Half-hour firecheck doors
L20.050	One-hour firecheck doors
L20.105	**Internal doors**
L20.110	Hardboard-faced flush doors for paint finish
L20.115	Plywood-faced flush doors for paint finish
L20.120	Factory finished flush doors (shrink wrapped)
L20.125	Half-hour fire resisting flush doors (30/30); hardboard-faced
L20.130	Half-hour fire resisting flush doors (30/30); plywood-faced
L20.135	Half-hour fire resisting flush doors (30/30); factory finished
L20.140	Panelled doors, softwood
L20.145	Panelled doors, hardwood

L20.205   **Door frames and door lining sets; wrought softwood**

L20.210   Frames; jambs or heads

L20.215   Frames; jambs or heads; once rebated

L20.220   Frames; jambs or heads; once rebated and grooved

L20.225   Frames; mullions or transoms

L20.230   Frames; twice rebated; mullions or transoms

L20.235   Frames; once sunk weathered; once rebated; three times grooved; sills

L20.240   Linings; tongued at angles

L20.245   Linings; once rebated; tongued at angles

L20.250   **Composite door frames; wrought softwood; assembled**

L20.255   Frames without sills; with iron dowels; opening in or out

L20.260   Frames with sills; opening in

L20.265   Frames with sills; opening out

L20.270   Firecheck frames with 25 mm rebates

L20.305   **Composite vestibule frames; wrought softwood; assembled**

L20.310   Vestibule frames for 838 × 1981 mm doors; opening in

L20.315   French frames for 1168 × 1981 mm doors; opening out

L20.320   **Composite garage door frames; wrought softwood; unassembled**

L20.325   Square section frames for up-and-over doors

L20.330   Rebated frames for side-hung doors

L20   TIMBER DOORS

	Labour (hr)				Doors (no.)	Unit

L20.005   **External doors**

L20.010   Panelled doors; hardwood; 44 mm thick

	Labour (hr)				Doors (no.)	Unit
838 × 1981 mm	1.38	—	—	—	1.025	no.

L20.015   Panelled doors; softwood; 44 mm thick

838 × 1981 mm	0.92	—	—	—	1.025	no.
762 × 1981 mm	0.92	—	—	—	1.025	no.

L20.020   Casement doors (pairs); 44 mm thick

1106 × 1994 mm	0.92	—	—	—	1.025	no.
1219 × 1981 mm	0.92	—	—	—	1.025	no.

L20.025   Framed ledged and braced doors; 44 mm thick

838 × 1981 mm	1.17	—	—	—	1.025	no.
762 × 1981 mm	1.17	—	—	—	1.025	no.

L20.030   Ledged and braced doors; 38 mm thick

838 × 1981 mm	1.00	—	—	—	1.025	no.
762 × 1981 mm	1.00	—	—	—	1.025	no.

L20.035   Stable doors; 44 mm thick

838 × 1981 mm	1.38	—	—	—	1.025	no.
762 × 1981 mm	1.38	—	—	—	1.025	no.

L20.040   Plywood faced flush doors; 44 mm thick

838 × 1981 mm	1.00	—	—	—	1.025	no.
762 × 1981 mm	1.00	—	—	—	1.025	no.

L20.045   Half-hour firecheck doors; 44 mm thick

838 × 1981 mm	1.38	—	—	—	1.025	no.
762 × 1981 mm	1.38	—	—	—	1.025	no.

L20.050   One-hour firecheck doors; 54 mm thick

838 × 1981 mm	1.70	—	—	—	1.025	no.
762 × 1981 mm	1.70	—	—	—	1.025	no.

	Labour (hr)				Doors (no.)	Unit
L20.105	**Internal doors**					
L20.110	Hardboard-faced, flush doors for paint finish; 35 mm thick					
838 × 1981 mm	0.92	—	—	—	1.025	no.
762 × 1981 mm	0.92	—	—	—	1.025	no.
686 × 1981 mm	0.92	—	—	—	1.025	no.
610 × 1981 mm	0.92	—	—	—	1.025	no.
533 × 1981 mm	0.92	—	—	—	1.025	no.
457 × 1981 mm	0.92	—	—	—	1.025	no.
L20.115	Plywood-faced, flush doors for paint finish; 35 mm thick					
762 × 1981 mm	0.87	—	—	—	1.025	no.
686 × 1981 mm	0.87	—	—	—	1.025	no.
610 × 1981 mm	0.87	—	—	—	1.025	no.
533 × 1981 mm	0.87	—	—	—	1.025	no.
457 × 1981 mm	0.87	—	—	—	1.025	no.
L20.120	Factory finished flush doors (shrink wrapped); 35 mm thick					
762 × 1981 mm	1.00	—	—	—	1.050	no.
686 × 1981 mm	1.00	—	—	—	1.050	no.
610 × 1981 mm	1.00	—	—	—	1.050	no.
533 × 1981 mm	1.00	—	—	—	1.050	no.
457 × 1981 mm	1.00	—	—	—	1.050	no.
L20.125	Half-hour fire resisting flush doors (30/30); hardboard faced for paint finish; 44 mm thick					
838 × 1981 mm	1.38	—	—	—	1.025	no.
762 × 1981 mm	1.38	—	—	—	1.025	no.
686 × 1981 mm	1.38	—	—	—	1.025	no.
L20.130	Half-hour fire resisting flush doors (30/30); plywood faced for paint finish; 44 mm thick					
838 × 1981 mm	1.31	—	—	—	1.025	no.
762 × 1981 mm	1.31	—	—	—	1.025	no.
686 × 1981 mm	1.31	—	—	—	1.025	no.
L20.135	Half-hour fire resisting flush doors (30/30); factory finished (shrink wrapped)					
838 × 1981 mm	1.50	—	—	—	1.050	no.
762 × 1981 mm	1.50	—	—	—	1.050	no.
686 × 1981 mm	1.50	—	—	—	1.050	no.

		Labour (hr)				Doors (no.)	Unit
L20.140	Panelled doors, softwood; 35 mm thick						
	838 × 1981 mm	0.92	—	—	—	1.025	no.
	762 × 1981 mm	0.92	—	—	—	1.025	no.
	686 × 1981 mm	0.92	—	—	—	1.025	no.
L20.145	Panelled doors, hardwood; 35 mm thick						
	762 × 1981 mm	1.38	—	—	—	1.025	no.
	686 × 1981 mm	1.38	—	—	—	1.025	no.
	610 × 1981 mm	1.38	—	—	—	1.025	no.

	Labour (hr)		Timber (ln. m)	Nails (kg)	Screws (no.)	Unit
L20.205	**Door frame and door lining sets; wrought softwood**					
L20.210	Frames; jambs or heads					
	32 × 63 mm	0.23	—	1.075	—	—	ln. m
	32 × 100 mm	0.23	—	1.075	—	—	ln. m
	32 × 150 mm	0.23	—	1.075	—	—	ln. m
	50 × 75 mm	0.25	—	1.075	—	—	ln. m
	50 × 100 mm	0.25	—	1.075	—	—	ln. m
	50 × 150 mm	0.25	—	1.075	—	—	ln. m
L20.215	Frames; jambs or heads; once rebated					
	50 × 75 mm	0.25	—	1.075	—	—	ln. m
	50 × 100 mm	0.25	—	1.075	—	—	ln. m
	50 × 150 mm	0.28	—	1.075	—	—	ln. m
	63 × 75 mm	0.28	—	1.075	—	—	ln. m
	63 × 100 mm	0.32	—	1.075	—	—	ln. m
	63 × 125 mm	0.32	—	1.075	—	—	ln. m
	63 × 150 mm	0.32	—	1.075	—	—	ln. m
L20.220	Frames; jambs or heads; once rebated; once grooved					
	50 × 100 mm	0.25	—	1.075	—	—	ln. m
	50 × 125 mm	0.25	—	1.075	—	—	ln. m
	50 × 150 mm	0.28	—	1.075	—	—	ln. m
	63 × 75 mm	0.32	—	1.075	—	—	ln. m
	63 × 100 mm	0.32	—	1.075	—	—	ln. m
	63 × 125 mm	0.32	—	1.075	—	—	ln. m
	63 × 150 mm	0.32	—	1.075	—	—	ln. m
L20.225	Frames; mullions or transoms					
	32 × 63 mm	0.17	—	1.075	—	—	ln. m
	32 × 100 mm	0.17	—	1.075	—	—	ln. m
	32 × 150 mm	0.17	—	1.075	—	—	ln. m
L20.230	Frames; twice rebated; mullions or transoms					
	38 × 125 mm	0.25	—	1.075	—	—	ln. m
	38 × 150 mm	0.25	—	1.075	—	—	ln. m
	50 × 100 mm	0.25	—	1.075	—	—	ln. m
	63 × 125 mm	0.32	—	1.075	—	—	ln. m
	63 × 150 mm	0.32	—	1.075	—	—	ln. m
L20.235	Frames; once sunk weathered; once rebated; three times grooved; sills					
	63 × 175 mm	0.32	—	1.075	—	—	ln. m
	75 × 150 mm	0.32	—	1.075	—	—	ln. m

		Labour (hr)		Timber (ln. m)	Nails (kg)	Screws (no.)	Unit
L20.240	Linings; tongued at angles						
	25 × 75 mm	0.23	—	1.075	—	—	ln. m
	25 × 100 mm	0.23	—	1.075	—	—	ln. m
	25 × 125 mm	0.23	—	1.075	—	—	ln. m
	25 × 150 mm	0.23	—	1.075	—	—	ln. m
	32 × 100 mm	0.23	—	1.075	—	—	ln. m
	32 × 125 mm	0.23	—	1.075	—	—	ln. m
	32 × 150 mm	0.23	—	1.075	—	—	ln. m
L20.245	Linings; once rebated; tongued at angles						
	38 × 100 mm	0.25	—	1.075	—	—	ln. m
	38 × 125 mm	0.25	—	1.075	—	—	ln. m
	38 × 150 mm	0.25	—	1.075	—	—	ln. m
L20.250	**Composite door frames; wrought softwood; assembled; hardwood sills, fitted with a waterbar; preservative treated**						
L20.255	Frames without sills; with iron dowels; opening in or out						
	FN 26, to suit door 762 × 1981 mm	0.58	—	1.00	—	—	no.
	FN 29, to suit door 838 × 1981 mm	0.58	—	1.00	—	—	no.
L20.260	Frames with sills; opening in						
	FNS 26, to suit door 762 × 1981 mm	0.67	—	1.00	—	—	no.
	FNS 29, to suit door 838 × 1981 mm	0.67	—	1.00	—	—	no.
L20.265	Frames with sills; opening out						
	FXS 26, to suit door 762 × 1981 mm	0.67	—	1.00	—	—	no.
	FXS 29, to suit door 838 × 1981 mm	0.67	—	1.00	—	—	no.
L20.270	Firecheck frames with 25 mm rebates, no waterbar						
	FCN 26, to suit door 762 × 1981 mm	0.67	—	1.00	—	—	no.
	FCN 29, to suit door 762 × 1981 mm	0.67	—	1.00	—	—	no.

	Labour (hr)		Frames (no.)	Nails (kg)	Screws (no.)	Unit
L20.305	**Composite vestibule frames; wrought softwood; assembled; fitted with a waterbar; hardwood sills**					
L20.310	Vestibule frames for 838 × 1981 mm doors opening in					
	1200 × 2083 mm	0.83	—	1.00	—	—	no.
	1350 × 2083 mm	0.83	—	1.00	—	—	no.
	1500 × 2083 mm	1.00	—	1.00	—	—	no.
	1800 × 2083 mm	1.00	—	1.00	—	—	no.
	2100 × 2083 mm	1.00	—	1.00	—	—	no.
L20.315	French frames for 1168 × 1981 mm doors, opening out					
	2172 × 2070 mm	1.00	—	1.00	—	—	no.
	2502 × 2070 mm	1.17	—	1.00	—	—	no.
L20.320	**Composite garage door frames; wrought softwood unassembled; preservative treated; iron dowels to jambs**					
L20.325	Square section frames for up-and-over doors					
	UOF 7066, to suit door 2134 × 1981 mm	1.75	—	1.00	—	—	no.
	UOF 7070, to suit door 2134 × 2134 mm	1.75	—	1.00	—	—	no.
L20.330	Rebated frames for side-hung doors					
	GDF 7070, to suit door 2134 × 2134 mm	1.75	—	1.00	—	—	no.

L21   *METAL DOORS*

L21.005   **Steel up-and-over garage doors; factory assembled**

L21.010   Galvanized finish

L21.015   **Patio doors; aluminium; factory double glazed**

L21.020   Two leaf doors; one fixed; one sliding

L21.025   Three leaf doors; one fixed; two sliding

L21   METAL DOORS

	Labour (hr)				Doors (no.)	Unit

L21.005 **Steel up and over garage doors; factory assembled with gear attached ready to install: overhead pre-tensioned spring operation, complete with guide frames and rollers; fixing with screws to timber subframe; (measured separately)**

		Labour (hr)				Doors (no.)	Unit
L21.010	Galvanized finish						
	2134 × 1981 mm	4.50	—	—	—	1.00	no.
	2134 × 2134 mm	5.25	—	—	—	1.00	no.
	2438 × 1981 mm	5.75	—	—	—	1.00	no.
	2438 × 2134 mm	6.50	—	—	—	1.00	no.

L21.015 **Patio doors; aluminium; factory double glazed with 18 mm toughened, double glazed units**

		Labour (hr)				Doors (no.)	Unit
L21.020	Two leaf doors; one fixed; one sliding						
	1800 mm wide × 2100 mm high	3.25	—	—	—	1.00	no.
	2100 mm wide × 2100 mm high	3.50	—	—	—	1.00	no.
L21.025	Three leaf doors; one fixed; two sliding						
	3000 mm wide × 2100 mm high	4.00	—	—	—	1.00	no.

L30   *TIMBER STAIRS/BALUSTRADES*

L30.005	**Standard staircase in wrought softwood**
L30.010	Straight flight stairs in one flight
L30.015	Extra for bullnosed step to stairs
L30.020	**Balustrades in wrought hemlock**
L30.025	Horizontal balustrades
L30.030	Raking balustrades
L30.035	**Balustrades in wrought Brazilian mahogany**
L30.040	Horizontal balustrades
L30.045	Raking balustrades
L30.050	Extra over handrail in wrought hemlock for horizontal turn
L30.055	80 × 80 mm newel posts in wrought hemlock
L30.060	Extra over handrail in wrought Brazilian mahogany for horizontal turn
L30.065	80 × 80 mm newel posts in wrought Brazilian mahogany

L30   TIMBER STAIRS/BALUSTRADES

	Labour (hr)	Stairs (no.)	Spindles (no.)	Handrail (ln. m)	String capping (ln. m)	Unit
L30.005 **Standard staircases in wrought softwood**						
L20.010 Straight flight stairs in one flight; 2676 mm going, 2600 mm rise; 12 treads; 13 risers						
864 mm wide	17.25	1.00	—	—	—	no.
914 mm wide	17.25	1.00	—	—	—	no.
L30.015 Extra for bullnosed step to stairs	0.58	1.00	—	—	—	no.
L30.020 **Balustrades in wrought hemlock; 75 × 75 mm moulded handrails; 41 × 41 mm turned balusters at 121 mm centres housed each end; 75 × 38 mm string capping**						
L30.025 Horizontal balustrades; overall height						
900 mm	3.25	—	8.33	1.075	1.075	ln. m
1000 mm	3.25	—	8.33	1.075	1.075	ln. m
L30.030 Raking balustrades; overall height						
900 mm	3.74	—	5.93	1.075	1.075	ln. m
1000 mm	3.74	—	5.93	1.075	1.075	ln. m
L30.035 **Balustrades in wrought Brazilian mahogany; 75 × 75 mm moulded handrails; 41 × 41 mm turned balusters at 121 mm centres housed each end; 75 × 38 mm string capping**						
L30.040 Horizontal balustrades; overall height						
900 mm	4.00	—	8.33	1.075	1.075	ln. m
1000 mm	4.00	—	8.33	1.075	1.075	ln. m
L30.045 Raking balustrades; overall height						
900 mm	4.60	—	5.93	1.075	1.075	ln. m
1000 mm	4.60	—	5.93	1.075	1.075	ln. m

		Labour (hr)	Horizontal turn (no.)	Newel base (no.)	Newel turning (no.)	Newel cap (no.)	Unit
L30.050	Extra over 75×75 mm handrail in wrought hemlock for						
	horizontal turn	0.33	1.00	—	—	—	no.
L30.055	80×80 mm newel posts in wrought hemlock						
	1375 mm long with turned cap	0.58	—	1.00	1.00	1.00	no.
	1570 mm long with turned cap	0.58	—	1.00	1.00	1.00	no.
L30.060	Extra over 75×75 mm handrail in wrought Brazilian mahogany for						
	horizontal turn	0.33	1.00	—	—	—	no.
L30.065	80×80 mm newel posts in wrought Brazilian mahogany						
	1375 mm long with turned cap	0.75	—	1.00	1.00	1.00	no.
	1570 mm long with turned cap	0.75	—	1.00	1.00	1.00	no.

L40 *GENERAL GLAZING*

L40.005	**Standard plain glass**
L40.010	3 mm glass to wood rebates with putty and sprigs
L40.015	4 mm glass to wood rebates with putty and sprigs
L40.020	5 mm glass to wood rebates with putty and sprigs
L40.025	6 mm glass to wood rebates with putty and sprigs
L40.030	3 mm glass to wood rebates with putty and bradded wood beads
L40.035	4 mm glass to wood rebates with putty and bradded wood beads
L40.040	5 mm glass to wood rebates with putty and bradded wood beads
L40.045	6 mm glass to wood rebates with putty and bradded wood beads
L40.050	3 mm glass to wood rebates with putty and loose screwed beads
L40.055	4 mm glass to wood rebates with putty and loose screwed beads
L40.060	5 mm glass to wood rebates with putty and loose screwed beads
L40.065	6 mm glass to wood rebates with putty and loose screwed beads

L40   GENERAL GLAZING

		Labour (hr)			Glass (sq. m)	Putty (kg)	Unit
L40.005	**Standard plain glass**						
L40.010	3 mm glass to wood rebates with putty and sprigs, in panes						
	not exceeding 0.15 sq. m	0.65	—	—	1.05	3.2	sq. m
	0.15–0.50 sq. m	0.52	—	—	1.05	2.0	sq. m
	0.50–1.00 sq. m	0.49	—	—	1.05	1.0	sq. m
	over 1.00 sq. m	0.46	—	—	1.05	0.5	sq. m
L40.015	4 mm glass to wood rebates with putty and sprigs, in panes						
	not exceeding 0.15 sq. m	0.68	—	—	1.05	3.2	sq. m
	0.15–0.50 sq. m	0.55	—	—	1.05	2.0	sq. m
	0.50–1.00 sq. m	0.51	—	—	1.05	1.0	sq. m
	over 1.00 sq. m	0.48	—	—	1.05	0.5	sq. m
L40.020	5 mm glass to wood rebates with putty and sprigs, in panes						
	not exceeding 0.15 sq. m	0.72	—	—	1.05	3.2	sq. m
	0.15–0.50 sq. m	0.57	—	—	1.05	2.0	sq. m
	0.50–1.00 sq. m	0.54	—	—	1.05	1.0	sq. m
	over 1.00 sq. m	0.51	—	—	1.05	0.5	sq. m
L40.025	6 mm glass to wood rebates with putty and sprigs, in panes						
	not exceeding 0.15 sq. m	0.75	—	—	1.05	3.2	sq. m
	0.15–0.50 sq. m	0.60	—	—	1.05	2.0	sq. m
	0.50–1.00 sq. m	0.56	—	—	1.05	1.0	sq. m
	over 1.00 sq. m	0.53	—	—	1.05	0.5	sq. m
L40.030	3 mm glass to wood rebates with putty and bradded wood beads, in panes						
	not exceeding 0.15 sq. m	0.75	—	—	1.05	1.0	sq. m
	0.15–0.50 sq. m	0.60	—	—	1.05	0.6	sq. m
	0.50–1.00 sq. m	0.56	—	—	1.05	0.3	sq. m
	over 1.00 sq. m	0.53	—	—	1.05	0.2	sq. m
L40.035	4 mm glass to wood rebates with putty and bradded wood beads, in panes						
	not exceeding 0.15 sq. m	0.79	—	—	1.05	1.0	sq. m
	0.15–0.50 sq. m	0.63	—	—	1.05	0.6	sq. m
	0.50–1.00 sq. m	0.59	—	—	1.05	0.3	sq. m
	over 1.00 sq. m	0.56	—	—	1.05	0.2	sq. m

		Labour (hr)			Glass (sq. m)	Putty (kg)	Unit
L40.040	5 mm glass to wood rebates with putty and bradded wood beads, in panes						
	not exceeding 0.15 sq. m	0.83	—	—	1.05	1.0	sq. m
	0.15–0.50 sq. m	0.66	—	—	1.05	0.6	sq. m
	0.50–1.00 sq. m	0.62	—	—	1.05	0.3	sq. m
	over 1.00 sq. m	0.59	—	—	1.05	0.2	sq. m
L40.045	6 mm glass to wood rebates with putty and bradded wood beads, in panes						
	not exceeding 0.15 sq. m	0.87	—	—	1.05	1.0	sq. m
	0.15–0.50 sq. m	0.69	—	—	1.05	0.6	sq. m
	0.50–1.00 sq. m	0.65	—	—	1.05	0.3	sq. m
	over 1.00 sq. m	0.62	—	—	1.05	0.2	sq. m
L40.050	3 mm glass to wood rebates with putty and loose screwed beads, in panes						
	not exceeding 0.15 sq. m	0.93	—	—	1.05	1.0	sq. m
	0.15–0.50 sq. m	0.74	—	—	1.05	0.6	sq. m
	0.50–1.00 sq. m	0.70	—	—	1.05	0.3	sq. m
	over 1.00 sq. m	0.66	—	—	1.05	0.2	sq. m
L40.055	4 mm glass to wood rebates with putty and loose screwed beads, in panes						
	not exceeding 0.15 sq. m	0.98	—	—	1.05	1.0	sq. m
	0.15–0.50 sq. m	0.78	—	—	1.05	0.6	sq. m
	0.50–1.00 sq. m	0.74	—	—	1.05	0.3	sq. m
	over 1.00 sq. m	0.69	—	—	1.05	0.2	sq. m
L40.060	5 mm glass to wood rebates with putty and loose screwed beads, in panes						
	not exceeding 0.15 sq. m	1.04	—	—	1.05	1.0	sq. m
	0.15–0.50 sq. m	0.83	—	—	1.05	0.6	sq. m
	0.50–1.00 sq. m	0.78	—	—	1.05	0.3	sq. m
	over 1.00 sq. m	0.73	—	—	1.05	0.2	sq. m
L40.065	6 mm glass to wood rebates with putty and loose screwed beads, in panes						
	not exceeding 0.15 sq. m	1.09	—	—	1.05	1.0	sq. m
	0.15–0.50 sq. m	0.87	—	—	1.05	0.6	sq. m
	0.50–1.00 sq. m	0.82	—	—	1.05	0.3	sq. m
	over 1.00 sq. m	0.77	—	—	1.05	0.2	sq. m

M SURFACE FINISHES

MA	**Notes**
1	The labour constants in Sections M10, and M20 are based on two craftsman plasterers and one labourer (two and one plastering gang).
2	The labour constants in Section M40 are based on one craftsman tiler.
3	The labour constants in Section M60 are based on one craftsman painter.

M10 *SAND CEMENT/GRANOLITHIC SCREEDS*

M10.005	**Sand cement screeds**
M10.010	To floors; level or to falls; trowelled
M10.015	To treads; level or to falls; trowelled
M10.020	To risers; trowelled
M10.025	To undercut risers; trowelled
M10.105	**Granolithic screeds**
M10.110	To floors; level or to falls; trowelled
M10.115	To treads; level or to falls; trowelled
M10.120	To risers; trowelled
M10.125	To undercut risers; trowelled
M10.130	19 mm thick skirting with rounded top edge and cove to paving

M10   SAND CEMENT/GRANOLITHIC SCREEDS

	Labour (hr)				Cement mortar (cu. m)	Unit
M10.005	**Sand cement screeds, internal**					
M10.010	To floors; level or to falls; trowelled					
over 300 mm wide						
25 mm thick	0.50	—	—	—	0.025	sq. m
32 mm thick	0.53	—	—	—	0.032	sq. m
38 mm thick	0.56	—	—	—	0.038	sq. m
50 mm thick	0.59	—	—	—	0.050	sq. m
65 mm thick	0.63	—	—	—	0.065	sq. m
75 mm thick	0.67	—	—	—	0.075	sq. m
not exceeding 300 mm wide						
25 mm thick	1.00	—	—	—	0.025	sq. m
32 mm thick	1.05	—	—	—	0.032	sq. m
38 mm thick	1.11	—	—	—	0.038	sq. m
50 mm thick	1.18	—	—	—	0.050	sq. m
65 mm thick	1.25	—	—	—	0.065	sq. m
75 mm thick	1.33	—	—	—	0.075	sq. m
M10.015	To treads; level or to falls; trowelled; 250 mm wide					
25 mm thick	0.42	—	—	—	0.006	ln. m
32 mm thick	0.44	—	—	—	0.008	ln. m
38 mm thick	0.46	—	—	—	0.010	ln. m
50 mm thick	0.49	—	—	—	0.013	ln. m
65 mm thick	0.52	—	—	—	0.016	ln. m
75 mm thick	0.56	—	—	—	0.019	ln. m
M10.020	To risers; trowelled; 175 mm wide					
19 mm thick	0.42	—	—	—	0.003	ln. m
25 mm thick	0.44	—	—	—	0.004	ln. m
M10.025	To undercut risers; trowelled; 175 mm wide					
19 mm thick	0.54	—	—	—	0.003	ln. m
25 mm thick	0.60	—	—	—	0.004	ln. m

	Labour (hr)				Granolithic (cu. m)	Unit
M10.105	**Granolithic screeds, cement and whinstone chippings, 6 mm down to dust (1:2.5); dust not to exceed 3 per cent passing BS.200 sieve**					
M10.110	To floors; level or to falls; trowelled					
	over 300 mm wide					
	12 mm thick laid monolithic	0.43	—	—	—	0.012	sq. m
	25 mm thick	0.60	—	—	—	0.025	sq. m
	32 mm thick	0.63	—	—	—	0.032	sq. m
	38 mm thick	0.67	—	—	—	0.038	sq. m
	50 mm thick	0.71	—	—	—	0.050	sq. m
	65 mm thick	0.75	—	—	—	0.065	sq. m
	75 mm thick	0.80	—	—	—	0.075	sq. m
	not exceeding 300 mm wide					
	12 mm thick laid monolithic	0.85	—	—	—	0.012	sq. m
	25 mm thick	1.20	—	—	—	0.025	sq. m
	32 mm thick	1.25	—	—	—	0.032	sq. m
	38 mm thick	1.33	—	—	—	0.038	sq. m
	50 mm thick	1.43	—	—	—	0.050	sq. m
	65 mm thick	1.50	—	—	—	0.065	sq. m
	75 mm thick	1.60	—	—	—	0.075	sq. m
M10.115	To treads; level or to falls; trowelled; 250 mm wide					
	25 mm thick	0.49	—	—	—	0.006	ln. m
	32 mm thick	0.52	—	—	—	0.008	ln. m
	38 mm thick	0.56	—	—	—	0.010	ln. m
	50 mm thick	0.57	—	—	—	0.013	ln. m
	65 mm thick	0.62	—	—	—	0.016	ln. m
	75 mm thick	0.64	—	—	—	0.019	ln. m
M10.120	To risers; trowelled; 175 mm wide					
	19 mm thick	0.49	—	—	—	0.003	ln. m
	25 mm thick	0.52	—	—	—	0.004	ln. m
M10.125	To undercut risers; trowelled; 175 mm wide					
	19 mm thick	0.60	—	—	—	0.003	ln. m
	25 mm thick	0.66	—	—	—	0.004	ln. m
M10.130	19 mm thick skirting with rounded top edge and cove to paving					
	150 mm high	0.86	—	—	—	0.003	ln. m
	225 mm high	0.94	—	—	—	0.004	ln. m
	ends	0.13	—	—		—	no.
	angles	0.15	—	—		—	no.

M20   *PLASTERED/RENDERED/COATINGS*

M20.005	**Plaster; Thistle; finishing coat of board finish**
M20.010	2 mm work to walls on plasterboard base
M20.015	2 mm work to ceilings on plasterboard base
M20.020	2 mm work to sides, soffits and tops of isolated beams on plasterboard base
M20.025	2 mm work to sides of isolated columns on plasterboard base
M20.030	**Plaster; Thistle; finishing coats of board finish**
M20.035	5 mm work to walls on plasterboard base
M20.040	5 mm work to ceilings on plasterboard base
M20.045	5 mm work to sides, soffits and tops of isolated beams on plasterboard base
M20.050	5 mm work to sides of isolated columns on plasterboard base
M20.105	**Plaster; Carlite; pre-mixed; floating coat of bonding; finish coat of finish**
M20.110	10 mm work to walls on plasterboard base
M20.115	10 mm work to ceilings on plasterboard base
M20.120	10 mm work to sides, soffits and tops of isolated beams on plasterboard base
M20.125	10 mm work to sides of isolated columns on plasterboard base
M20.205	**Plaster; Carlite; pre-mixed; floating coat of browning; finish coat of finish**
M20.210	13 mm work to walls on brickwork or blockwork base
M20.215	13 mm work to sides of isolated columns on brickwork or blockwork base

M20.220	**Plaster; Carlite; pre-mixed; floating coat of bonding; finish coat of finish**
M20.225	10 mm work to walls on concrete base
M20.230	10 mm work to ceilings on concrete base
M20.235	10 mm work to sides, soffits and tops of isolated beams on concrete base
M20.240	10 mm work to sides of isolated columns on concrete base
M20.305	**Plaster; Carlite; pre-mixed; floating coat of lathing; finish coat of finish**
M20.310	13 mm work to walls on metal lathing base
M20.315	13 mm work to ceilings on metal lathing base
M20.320	13 mm work to sides, soffits and tops of isolated beams on metal lathing base
M20.325	13 mm work to sides of isolated columns on metal lathing base
M20.405	**Gypsum plaster core cove cornice; fixed with adhesive**
M20.410	100 mm girth
M20.415	127 mm girth
M20.420	**Accessories; internal**
M20.425	Angle beads; galvanized steel; fixing with plasterdabs
M20.430	Stop beads; galvanized steel; fixing with plaster dabs
M20.435	Depth gauge beads; galvanized steel; fixing with plaster dabs
M20.440	Movement beads; galvanized steel; fixing with plaster dabs
M20.445	Architrave beads; galvanized steel; fixing with plaster dabs

M20.505	**Mortar, cement and sand; internal**
M20.510	To walls on brickwork or blockwork base; trowelled
M20.515	To sides of isolated columns on brickwork or blockwork base
M20.520	Rounded external angle
M20.525	Rounded internal angle
M20.530	**Accessories; external**
M20.535	Angle beads; fixing with mortar dabs
M20.540	Stop beads; fixing with mortar dabs
M20.545	Movement beads; fixing with mortar dabs

M20   PLASTERED/RENDERED/COATINGS

		Labour (hr)			Plaster undercoat (tonne)	Plaster finish (tonne)	Unit
M20.005	**Plaster; Thistle; finishing coat of board finish; 2 mm thick, one coat work; steel trowelled; internal**						
M20.010	2 mm work to walls on plasterboard base						
	over 300 mm wide	0.20	—	—	—	0.0025	sq. m
	not exceeding 300 mm wide	0.40	—	—	—	0.0025	sq. m
M20.015	2 mm work to ceilings on plasterboard base						
	over 300 mm wide	0.29	—	—	—	0.0025	sq. m
	not exceeding 300 mm wide	0.58	—	—	—	0.0025	sq. m
M20.020	2 mm work to sides, soffits and tops of isolated beams on plasterboard base						
	over 300 mm wide	0.33	—	—	—	0.0025	sq. m
	not exceeding 300 mm wide	0.66	—	—	—	0.0025	sq. m
M20.025	2 mm work to sides of isolated columns on plasterboard base						
	over 300 mm wide	0.25	—	—	—	0.0025	sq. m
	not exceeding 300 mm wide	0.50	—	—	—	0.0025	sq. m
M20.030	**Plaster, Thistle; finishing coats of board finish, 5 mm thick, two coat work; steel trowelled; internal**						
M20.035	5 mm work to walls on plasterboard base						
	over 300 mm wide	0.30	—	—	—	0.0061	sq. m
	not exceeding 300 mm wide	0.60	—	—	—	0.0061	sq. m
M20.040	5 mm work to ceilings on plasterboard base						
	over 300 mm wide	0.43	—	—	—	0.0061	sq. m
	not exceeding 300 mm wide	0.86	—	—	—	0.0061	sq. m
M20.045	5 mm work to sides, soffits and tops of isolated beams on plasterboard base						
	over 300 mm wide	0.50	—	—	—	0.0061	sq. m
	not exceeding 300 mm wide	1.00	—	—	—	0.0061	sq. m
M20.050	5 mm work to sides of isolated columns on plasterboard base						
	over 300 mm wide	0.38	—	—	—	0.0061	sq. m
	not exceeding 300 mm wide	0.76	—	—	—	0.0061	sq. m

		Labour (hr)			Plaster undercoat (tonne)	Plaster finish (tonne)	Unit
M20.105	**Plaster; Carlite; pre-mixed; floating coat of bonding, 8 mm thick; finish coat of finish, 2 mm thick; steel trowelled; internal**						
M20.110	10 mm work to walls on plasterboard base						
	over 300 mm wide	0.55	—	—	0.0063	0.0022	sq. m
	not exceeding 300 mm wide	1.10	—	—	0.0063	0.0022	sq. m
M20.115	10 mm work to ceilings on plasterboard base						
	over 300 mm wide	0.78	—	—	0.0063	0.0022	sq. m
	not exceeding 300 mm wide	1.56	—	—	0.0063	0.0022	sq. m
M20.120	10 mm work to sides, soffits and tops of isolated beams on plasterboard base						
	over 300 mm wide	0.91	—	—	0.0063	0.0022	sq. m
	not exceeding 300 mm wide	1.82	—	—	0.0063	0.0022	sq. m
M20.125	10 mm work to sides of isolated columns on plasterboard base						
	over 300 mm wide	0.68	—	—	0.0063	0.0022	sq. m
	not exceeding 300 mm wide	1.36	—	—	0.0063	0.0022	sq. m

		Undercoat				
	Labour (hr)	Browning (tonne)	Bonding (tonne)	Lathing (tonne)	Finish (tonne)	Unit
M20.205 **Plaster, Carlite; pre-mixed; floating coat of browning, 11 mm thick; finish coat of finish, 2 mm thick; steel trowelled; internal**						
M20.210 13 mm work to walls on brickwork or blockwork base						
over 300 mm wide	0.46	0.0071	—	—	0.0022	sq. m
not exceeding 300 mm wide	0.92	0.0071	—	—	0.0022	sq. m
M20.215 13 mm work to sides of isolated columns on brickwork or blockwork base						
over 300 mm wide	0.58	0.0071	—	—	0.0022	sq. m
not exceeding 300 mm wide	1.16	0.0071	—	—	0.0022	sq. m
M20.220 **Plaster, Carlite; pre-mixed; floating coat of bonding, 8 mm thick; finish coat of finish plaster, 2 mm thick; steel trowelled; internal**						
M20.225 10 mm work to walls on concrete base						
over 300 mm wide	0.35	—	0.0067	—	0.0022	sq. m
not exceeding 300 mm wide	0.70	—	0.0067	—	0.0022	sq. m
M20.230 10 mm work to ceilings on concrete base						
over 300 mm wide	0.50	—	0.0067	—	0.0022	sq. m
not exceeding 300 mm wide	1.00	—	0.0067	—	0.0022	sq. m
M20.235 10 mm work to sides, soffits and tops of isolated beams on concrete base						
over 300 mm wide	0.59	—	0.0067	—	0.0022	sq. m
not exceeding 300 mm wide	1.18	—	0.0067	—	0.0022	sq. m
M20.240 10 mm work to sides of isolated columns on concrete base						
over 300 mm wide	0.44	—	0.0067	—	0.0022	sq. m
not exceeding 300 mm wide	0.88	—	0.0067	—	0.0022	sq. m

| | Labour (hr) | Undercoat | | | | Unit |
		Browning (tonne)	Bonding (tonne)	Lathing (tonne)	Finish (tonne)	
M20.305 **Plaster, Carlite; pre-mixed; floating coat of lathing, 11 mm thick; finish coat of finish, 2 mm thick; steel trowelled; internal**						
M20.310 13 mm work to walls on metal lathing base						
over 300 mm wide	0.55	—	—	0.0154	0.0022	sq. m
not exceeding 300 mm wide	1.10	—	—	0.0154	0.0022	sq. m
M20.315 13 mm work to ceilings on metal lathing base						
over 300 mm wide	0.78	—	—	0.0154	0.0022	sq. m
not exceeding 300 mm wide	1.56	—	—	0.0154	0.0022	sq. m
M20.320 13 mm work to sides, soffits and tops of isolated beams on metal lathing base						
over 300 mm wide	0.91	—	—	0.0154	0.0022	sq. m
not exceeding 300 mm wide	1.82	—	—	0.0154	0.0022	sq. m
M20.325 13 mm work to sides of isolated columns on metal lathing base						
over 300 mm wide	0.68	—	—	0.0154	0.0022	sq. m
not exceeding 300 mm wide	1.36	—	—	0.0154	0.0022	sq. m

		Labour (hr)		Cove (ln. m)	Adhesive (kg)	Beads (ln. m)	Unit
M20.405	**Gypsum plaster core cove cornice, fixed with adhesive**						
M20.410	100 mm girth	0.18	—	1.05	0.18	—	ln. m
	returned end	0.25	—	—	—	—	no.
	angles	0.33	—	—	—	—	no.
M20.415	127 mm girth	0.20	—	1.05	0.31	—	ln. m
	returned ends	0.28	—	—	—	—	no.
	angles	0.42	—	—	—	—	no.
M20.420	**Accessories; internal**						
M20.425	Angle beads; galvanized steel; fixing with plaster dabs to brickwork or blockwork; Expamet; reference						
	Ref. 550; standard	0.14	—	—	—	1.05	ln. m
	Ref. 558; maxicon	0.14	—	—	—	1.05	ln. m
	Ref. 559; square nose	0.14	—	—	—	1.05	ln. m
M20.430	Stop beads; galvanized steel; fixing with plaster dabs to brickwork or blockwork; Expamet; reference						
	Ref. 562; 10 mm deep	0.13	—	—	—	1.05	ln. m
	Ref. 563; 13 mm deep	0.13	—	—	—	1.05	ln. m
	Ref. 565; 16 mm deep	0.13	—	—	—	1.05	ln. m
	Ref. 566; 19 mm deep	0.13	—	—	—	1.05	ln. m
M20.435	Depth gauge beads; galvanized steel; fixing with plaster dabs to brickwork or blockwork; Expamet; reference						
	Ref. 569	0.13	—	—	—	1.05	ln. m
M20.440	Movement beads; galvanized steel; fixing with plaster dabs to brickwork or blockwork; Expamet; reference						
	Ref. 588; 10 mm deep	0.16	—	—	—	1.05	ln. m
M20.445	Architrave beads; galvanized steel; fixing with plaster dabs to brickwork or blockwork; Expamet; reference						
	Ref. 579; 13 mm deep	0.14	—	—	—	1.05	ln. m
	Ref. 580; 13 mm deep	0.14	—	—	—	1.05	ln. m
	Ref. 585; 10 mm deep	0.14	—	—	—	1.05	ln. m
	Ref. 586; 10 mm deep	0.14	—	—	—	1.05	ln. m

	Labour (hr)			Cement mortar (cu. m)	Beads (ln. m)	Unit
M20.505	**Mortar, cement and sand; internal**					
M20.510	To walls on brickwork or blockwork base; trowelled					
	12 mm thick; one coat work					
	over 300 mm wide	0.40	—	—	0.012	—	sq. m
	not exceeding 300 mm wide	0.80	—	—	0.012	—	sq. m
	19 mm thick; two coat work					
	over 300 mm wide	0.60	—	—	0.019	—	sq. m
	not exceeding 300 mm wide	1.20	—	—	0.019	—	sq. m
M20.515	To sides of isolated columns on brickwork or blockwork base; trowelled					
	12 mm thick; one coat work					
	over 300 mm wide	0.75	—	—	0.012	—	sq. m
	not exceeding 300 mm wide	1.50	—	—	0.012	—	sq. m
	19 mm thick; two coat work					
	over 300 mm wide	1.25	—	—	0.019	—	sq. m
	not exceeding 300 mm wide	2.50	—	—	0.019	—	sq. m
M20.520	Rounded external angles					
	10–100 mm radius	0.25	—	—	—	—	ln. m
M20.525	Rounded internal angles					
	10–100 mm radius	0.30	—	—	—	—	ln. m
M20.530	**Accessories; external**					
M20.535	Angle beads; fixing with mortar dabs to brickwork or blockwork; Expamet; reference					
	Ref. 1019	0.16	—	—	—	1.05	ln. m
M20.540	Stop beads; fixing with mortar dabs to brickwork or blockwork; Expamet; reference					
	Ref. 1222	0.15	—	—	—	1.05	ln. m
	Ref. 1229; Type 2	0.15	—	—	—	1.05	ln. m
	Ref. 570	0.15	—	—	—	1.05	ln. m
M20.545	Movement beads; galvanised steel; fixing with mortar dabs to brickwork or blockwork; Expamet; reference					
	Ref. 590 19 mm deep	0.16	—	—	—	1.05	ln. m

M40   *QUARRY/CERAMIC TILING*

M40.005	**Quarry tiles; bedding in cement mortar**
M40.010	To floors on concrete base
M40.015	To treads on concrete base
M40.020	To risers on concrete base
M40.025	Raking cutting
M40.030	Curved cutting
M40.035	Cut and fit tiles around steel joists, angles, pipes or the like
M40.040	Skirting, including 13 mm bed of mortar
M40.105	**Ceramic tiles, glazed; fixing with Bal-Wall adhesive; grouting with Bal-Grout**
M40.110	To walls on plaster base
M40.115	Raking cutting
M40.120	Curved cutting
M40.125	Cut and fit around steel joists, angles, pipes or the like

M40.205	**Ceramic tiles, unglazed; bedding in cement mortar**
M40.210	To floors on concrete base
M40.215	To landings on concrete base
M40.220	To treads on concrete base
M40.225	To risers on concrete base
M40.230	Raking cutting
M40.235	Curved cutting
M40.240	Cut and fit around steel joists, angles, pipes or the like
M40.245	Skirting, including 13 mm bed of cement mortar

M40   QUARRY/CERAMIC TILING

		Labour (hr)			Tiles (no.)	Cement mortar (cu. m)	Unit
M40.005	**Quarry tiles; 6 mm joints; symmetrical layout; bedding in 13 mm cement mortar (1:3); pointing with cement mortar (1:1)**						
M40.010	To floors on concrete base						
	over 300 mm wide						
	152 × 152 × 16 mm	1.95	—	—	42.06	0.013	sq. m
	228 × 228 × 32 mm	1.43	—	—	19.18	0.013	sq. m
	not exceeding 300 mm wide						
	152 × 152 × 16 mm	3.90	—	—	48.07	0.013	sq. m
	228 × 228 × 32 mm	2.85	—	—	21.92	0.013	sq. m
M40.015	To treads on concrete base; 250 mm wide						
	152 × 152 × 16 mm	0.98	—	—	11.52	0.003	ln. m
	228 × 228 × 32 mm	0.72	—	—	5.25	0.003	ln. m
M40.020	To risers on concrete base; 175 mm wide						
	152 × 152 × 16 mm	0.75	—	—	8.06	0.002	ln. m
	228 × 228 × 32 mm	0.53	—	—	3.68	0.002	ln. m
M40.025	Raking cutting						
	152 × 152 × 16 mm	0.17	—	—	2.10	—	ln. m
	228 × 228 × 32 mm	0.20	—	—	0.96	—	ln. m
M40.030	Curved cutting						
	152 × 152 × 16 mm	0.33	—	—	2.94	—	ln. m
	228 × 228 × 32 mm	0.39	—	—	1.34	—	ln. m

		Labour (hr)			Tiles (no.)	Cement mortar (cu. m)	Unit
M40.035	Cut and fit tiles around steel joists, angles, pipes or the like						
	not exceeding 0.30 m girth						
	152 × 152 × 16 mm	0.12	—	—	1.05	—	no.
	228 × 228 × 32 mm	0.14	—	—	0.48	—	no.
	0.30–1.00 m girth						
	152 × 152 × 16 mm	0.20	—	—	2.10	—	no.
	228 × 228 × 32 mm	0.23	—	—	0.96	—	no.
	1.00–2.00 m girth						
	152 × 152 × 16 mm	0.38	—	—	3.79	—	no.
	228 × 228 × 32 mm	0.45	—	—	1.73	—	no.
	over 2.00 m girth						
	152 × 152 × 16 mm	0.26	—	—	2.10	—	ln. m
	228 × 228 × 32 mm	0.30	—	—	0.96	—	ln. m
M40.040	Skirting, including 13 mm bed of cement mortar (1:3), jointing and pointing with cement mortar (1:1)						
	152 × 152 coved base with paving	0.68	—	—	6.65	0.002	ln. m
	ends; fitted	0.08	—	—	—	—	no.
	internal angle	0.15	—	—	1.05	—	no.
	external angle	0.15	—	—	1.05	—	no.

		Labour (hr)		Tiles (no.)	Adhesive (litre)	Grout (kg)	Unit
M40.105	**Ceramic tiles, BS.1281, glazed; 2 mm joints, symmetrical layout; fixing with Bal-Wall adhesive; grouting with Bal-Grout; internal**						
M40.110	To walls on plaster base						
	over 300 mm wide						
	108 × 108 × 4 mm	2.90	—	86.78	1.00	0.28	sq. m
	152 × 152 × 6 mm	2.00	—	44.27	1.00	0.30	sq. m
	203 × 102 × 6 mm	2.20	—	49.25	1.00	0.33	sq. m
	not exceeding 300 mm wide						
	108 × 108 × 4 mm	5.80	—	95.04	1.00	0.28	sq. m
	152 × 152 × 6 mm	4.00	—	48.49	1.00	0.30	sq. m
	203 × 102 × 6 mm	4.40	—	53.94	1.00	0.33	sq. m
M40.115	Raking cutting						
	108 × 108 × 4 mm	0.12	—	4.34	—	—	ln. m
	152 × 152 × 6 mm	0.13	—	2.21	—	—	ln. m
	203 × 102 × 6 mm	0.15	—	2.46	—	—	ln. m
M40.120	Curved cutting						
	108 × 108 × 4 mm	0.24	—	6.07	—	—	ln. m
	152 × 152 × 6 mm	0.27	—	3.10	—	—	ln. m
	203 × 102 × 6 mm	0.30	—	3.45	—	—	ln. m
M40.125	Cut and fit tiles around steel joists, angles, pipes or the like						
	not exceeding 0.30 m girth						
	108 × 108 × 4 mm	0.08	—	1.74	—	—	no.
	152 × 152 × 6 mm	0.09	—	0.89	—	—	no.
	203 × 102 × 6 mm	0.11	—	0.99	—	—	no.
	0.30–1.00 m girth						
	108 × 108 × 4 mm	0.12	—	3.47	—	—	no.
	152 × 152 × 6 mm	0.15	—	1.77	—	—	no.
	203 × 102 × 6 mm	0.17	—	1.97	—	—	no.
	1.00–2.00 m girth						
	108 × 108 × 4 mm	0.27	—	6.94	—	—	no.
	152 × 152 × 6 mm	0.32	—	3.54	—	—	no.
	203 × 102 × 6 mm	0.35	—	3.94	—	—	no.
	over 2.00 m girth						
	108 × 108 × 4 mm	0.18	—	4.34	—	—	ln. m
	152 × 152 × 6 mm	0.21	—	2.21	—	—	ln. m
	203 × 102 × 6 mm	0.23	—	2.46	—	—	ln. m

		Labour (hr)			Tiles (no.)	Cement mortar (cu. m)	Unit
M40.205	**Ceramic tiles, unglazed; 6 mm joints, symmetrical layout; bedding in 13 mm cement mortar (1:3); pointing with cement mortar (1:1)**						
M40.210	To floors on concrete base						
	over 300 mm wide						
	100 × 100 × 9.5 mm	2.85	—	—	93.45	0.013	sq. m
	152 × 152 × 12.5 mm	1.80	—	—	42.06	0.013	sq. m
	250 × 125 × 9.5 mm	1.58	—	—	31.31	0.013	sq. m
	not exceeding 300 mm wide						
	100 × 100 × 9.5 mm	5.70	—	—	102.35	0.013	sq. m
	152 × 152 × 12.5 mm	3.60	—	—	46.07	0.013	sq. m
	250 × 125 × 9.5 mm	3.15	—	—	34.29	0.013	sq. m
M40.215	To landings on concrete base						
	over 300 mm wide						
	100 × 100 × 9.5 mm	4.28	—	—	93.45	0.013	sq. m
	152 × 152 × 12.5 mm	2.70	—	—	42.06	0.013	sq. m
	250 × 125 × 9.5 mm	2.37	—	—	31.31	0.013	sq. m
M40.220	To treads on concrete base						
	250 mm wide						
	100 × 100 × 9.5 mm	1.43	—	—	25.58	0.003	ln. m
	152 × 152 × 12.5 mm	0.90	—	—	11.52	0.003	ln. m
	250 × 125 × 9.5 mm	0.83	—	—	8.58	0.003	ln. m
M40.225	To risers on concrete base						
	175 mm wide						
	100 × 100 × 9.5 mm	1.00	—	—	17.91	0.002	ln. m
	152 × 152 × 12.5 mm	0.68	—	—	8.06	0.002	ln. m
	250 × 125 × 9.5 mm	0.56	—	—	6.00	0.002	ln. m
M40.230	Raking cutting						
	100 × 100 × 9.5 mm	0.12	—	—	4.67	—	ln. m
	152 × 152 × 12.5 mm	0.15	—	—	2.10	—	ln. m
	250 × 125 × 9.5 mm	0.12	—	—	1.57	—	ln. m
M40.235	Curved cutting						
	100 × 100 × 9.5 mm	0.24	—	—	6.54	—	ln. m
	152 × 152 × 12.5 mm	0.30	—	—	2.94	—	ln. m
	250 × 125 × 9.5 mm	0.24	—	—	2.19	—	ln. m

		Labour (hr)			Tiles (no.)	Cement mortar (cu. m)	Unit
M40.240	Cut and fit tiles around steel joists, angles, pipes or the like						
	not exceeding 0.30 m girth						
	100 × 100 × 9.5 mm	0.08	—	—	2.34	—	no.
	152 × 152 × 12.5 mm	0.11	—	—	1.05	—	no.
	250 × 125 × 9.5 mm	0.08	—	—	0.78	—	no.
	0.30–1.00 m girth						
	100 × 100 × 9.5 mm	0.14	—	—	3.74	—	no.
	152 × 152 × 12.5 mm	0.17	—	—	1.68	—	no.
	250 × 125 × 9.5 mm	0.14	—	—	1.25	—	no.
	1.00–2.00 m girth						
	100 × 100 × 9.5 mm	0.27	—	—	8.41	—	no.
	152 × 152 × 12.5 mm	0.35	—	—	3.79	—	no.
	250 × 125 × 9.5 mm	0.27	—	—	2.82	—	no.
	over 2.00 m girth						
	100 × 100 × 9.5 mm	0.18	—	—	5.61	—	ln. m
	152 × 152 × 12.5 mm	0.23	—	—	2.52	—	ln. m
	250 × 125 × 9.5 mm	0.18	—	—	1.88	—	ln. m
M40.245	Skirting, including 13 mm bed of cement mortar (1:3), jointing and pointing with cement mortar (1:1)						
	152 × 100 × 9.5 mm; butt joints; coved						
	junction with paving; 100 mm high	0.30	—	—	6.82	0.001	ln. m
	ends; fitted	0.08	—	—	—	—	no.
	internal angle	0.15	—	—	1.05	—	no.
	external angle	0.15	—	—	1.05	—	no.
	200 × 100 × 9.5 mm; butt joints; coved						
	junction with paving; 100 mm high	0.38	—	—	5.20	0.001	ln. m
	ends; fitted	0.08	—	—	—	—	no.
	internal angle	0.15	—	—	1.05	—	no.
	external angle	0.15	—	—	1.05	—	no.
	152 × 127 × 9.5 mm; butt joints; coved						
	junction with paving; 127 mm high	0.38	—	—	6.82	0.001	ln. m
	ends; fitted	0.08	—	—	—	—	no.
	internal angle	0.15	—	—	1.05	—	no.
	external angle	0.15	—	—	1.05	—	no.

M60   *PAINTING/CLEAR FINISHING*

M60.005	**One coat primer**
M60.010	Wood general surfaces
M60.015	Wood skirtings or the like
M60.020	Wood frames, linings or associated mouldings
M60.025	Cornices
M60.030	Staircases
M60.035	Wood partly glazed doors or screens
M60.040	Wood windows
M60.105	**One undercoat; one coat oil based full gloss finish**
M60.110	Wood general surfaces
M60.115	Wood skirtings or the like
M60.120	Wood frames, linings or associated mouldings
M60.125	Cornices
M60.130	Staircases
M60.135	Wood partly glazed doors or screens
M60.140	Wood windows

M60.205   **Sadolins Classic PX65 decorative wood protection**

M60.210   Wood general surfaces

M60.215   Wood skirtings or the like

M60.220   Wood frames, linings or associated mouldings

M60.225   Cornices

M60.230   Staircases

M60.235   Wood partly glazed doors or screens

M60.240   Wood windows

M60   PAINTING/CLEAR FINISHING

	Labour (hr)		Primer (litre)	Undercoat (litre)	Gloss (litre)	Unit
M60.005	**One coat primer**					
M60.010	Wood general surfaces					
0–150 mm girth	0.07	—	0.016	—	—	ln. m
150–300 mm girth	0.10	—	0.032	—	—	ln. m
over 300 mm girth	0.22	—	0.105	—	—	sq. m
isolated, not exceeding 0.50m²	0.07	—	0.053	—	—	no.
M60.015	Wood skirtings or the like					
0–150 mm girth	0.07	—	0.016	—	—	ln. m
150–300 mm girth	0.11	—	0.032	—	—	ln. m
over 300 mm girth	0.24	—	0.105	—	—	sq. m
M60.020	Wood frames, linings or associated mouldings					
0–150 mm girth	0.07	—	0.016	—	—	ln. m
150–300 mm girth	0.10	—	0.032	—	—	ln. m
over 300 mm girth	0.22	—	0.105	—	—	sq. m
M60.025	Cornices					
0–150 mm girth	0.08	—	0.016	—	—	ln. m
150–300 mm girth	0.12	—	0.032	—	—	ln. m
over 300 mm girth	0.26	—	0.105	—	—	sq. m
M60.030	Staircases					
0–150 mm girth	0.07	—	0.016	—	—	ln. m
150–300 mm girth	0.10	—	0.032	—	—	ln. m
over 300 mm girth	0.23	—	0.105	—	—	sq. m
M60.035	Wood partly glazed doors or screens; over 300 mm girth					
small panes (not exceeding 0.1 m²)	0.50	—	0.063	—	—	sq. m
medium panes (0.1–0.5 m²)	0.36	—	0.053	—	—	sq. m
large panes (0.5–1.0 m²)	0.33	—	0.042	—	—	sq. m
extra large panes (over 1.0 m²)	0.31	—	0.037	—	—	sq. m
M60.040	Wood windows over 300 mm girth					
small panes (not exceeding 0.1 m²)	0.55	—	0.084	—	—	sq. m
medium panes (0.10–0.5 m²)	0.38	—	0.074	—	—	sq. m
large panes (0.5–1.0 m²)	0.36	—	0.063	—	—	sq. m
extra large panes (over 1.0 m²)	0.33	—	0.053	—	—	sq. m
edges of opening casements	0.06	—	0.016	—	—	ln. m

	Labour (hr)	Primer (litre)	Undercoat (litre)	Gloss (litre)	Unit
M60.105 **One undercoat: one coat oil based full gloss finish**					
M60.110 Wood general surfaces					
0–150 mm girth	0.11	—	—	0.012	0.012	ln. m
150–300 mm girth	0.17	—	—	0.024	0.024	ln. m
over 300 mm girth	0.36	—	—	0.080	0.080	sq. m
isolated area, not exceeding 0.50 m^2	0.14	—	—	0.040	0.040	no.
M60.115 Wood skirtings or the like					
0–150 mm girth	0.12	—	—	0.012	0.012	ln. m
150–300 mm girth	0.17	—	—	0.024	0.024	ln. m
over 300 mm girth	0.39	—	—	0.080	0.080	sq. m
M60.120 Wood frames, linings or associated mouldings					
0–150 mm girth	0.11	—	—	0.012	0.012	ln. m
150–300 mm girth	0.17	—	—	0.024	0.024	ln. m
over 300 mm girth	0.36	—	—	0.080	0.080	sq. m
M60.125 Cornices					
0–150 mm girth	0.13	—	—	0.012	0.012	ln. m
150–300 mm girth	0.19	—	—	0.024	0.024	ln. m
over 300 mm girth	0.42	—	—	0.080	0.080	sq. m
M60.130 Staircases					
0–150 mm girth	0.11	—	—	0.012	0.012	ln. m
150–300 mm girth	0.17	—	—	0.024	0.024	ln. m
over 300 mm girth	0.37	—	—	0.080	0.080	sq. m
M60.135 Wood partly glazed doors or screens: over 300 mm girth					
small panes (not exceeding 0.1 m^2)	0.90	—	—	0.048	0.048	sq. m
medium panes (0.1–0.5 m^2)	0.64	—	—	0.040	0.040	sq. m
large panes (0.5–1.0 m^2)	0.58	—	—	0.033	0.033	sq. m
extra large panes (over 1.0 m^2)	0.53	—	—	0.028	0.028	sq. m
M60.140 Wood windows; over 300 mm girth					
small panes (not exceeding 0.1 m^2)	1.02	—	—	0.064	0.064	sq. m
medium panes (0.1–0.5 m^2)	0.68	—	—	0.056	0.056	sq. m
large panes (0.5–1.0 m^2)	0.64	—	—	0.048	0.048	sq. m
extra large panes (over 1.0 m^2)	0.58	—	—	0.040	0.040	sq. m
edges of opening casements	0.12	—	—	0.013	0.013	ln. m

	Labour (hr)				Sadolin PX65 (litre)	Unit
M60.205 **Sadolin Classic PX65 decorative wood protection**						
M60.210 General surfaces						
first coat to untreated surfaces						
0–150 mm girth	0.03	—	—	—	0.016	ln. m
150–300 mm girth	0.06	—	—	—	0.032	ln. m
over 300 mm girth	0.13	—	—	—	0.108	sq. m
isolated, not exceeding 0.5 m²	0.03	—	—	—	0.054	no.
second and subsequent coats						
0–150 mm girth	0.03	—	—	—	0.013	ln. m
150–300 mm girth	0.06	—	—	—	0.026	ln. m
over 300 mm girth	0.13	—	—	—	0.085	sq. m
isolated, not exceeding 0.5 m²	0.03	—	—	—	0.043	no.
M60.215 Wood skirtings or the like						
first coat to untreated surfaces						
0–150 mm girth	0.03	—	—	—	0.016	ln. m
150–300 mm girth	0.06	—	—	—	0.032	ln. m
over 300 mm girth	0.13	—	—	—	0.108	sq. m
second and subsequent coats						
0–150 mm girth	0.03	—	—	—	0.013	ln. m
150–300 mm girth	0.06	—	—	—	0.026	ln. m
over 300 mm girth	0.14	—	—	—	0.085	sq. m
M60.220 Wood frames, linings or associated mouldings						
first coat to untreated surfaces						
0–150 mm girth	0.03	—	—	—	0.016	ln. m
150–300 mm girth	0.06	—	—	—	0.032	ln. m
over 300 mm girth	0.13	—	—	—	0.108	sq. m
second and subsequent coats						
0–150 mm girth	0.03	—	—	—	0.013	ln. m
150–300 mm girth	0.06	—	—	—	0.026	ln. m
over 300 mm girth	0.15	—	—	—	0.085	sq. m
M60.225 Cornices;						
first coat on untreated surfaces						
0–150 mm girth	0.04	—	—	—	0.016	ln. m
150–300 mm girth	0.07	—	—	—	0.032	ln. m
over 300 mm girth	0.15	—	—	—	0.108	sq. m
second and subsequent coats						
0–150 mm girth	0.04	—	—	—	0.016	ln. m
150–300 mm girth	0.07	—	—	—	0.032	ln. m
over 300 mm girth	0.15	—	—	—	0.085	sq. m

		Labour (hr)				Sadolin PX65 (litre)	Unit
M60.230	Staircases						
	first coat to untreated surfaces						
	0–150 mm girth	0.03	—	—	—	0.016	ln. m
	150–300 mm girth	0.06	—	—	—	0.032	ln. m
	over 300 mm girth	0.13	—	—	—	0.108	sq. m
	second and subsequent coats						
	0–150 mm girth	0.03	—	—	—	0.013	ln. m
	150–300 mm girth	0.06	—	—	—	0.026	ln. m
	over 300 mm girth	0.13	—	—	—	0.085	sq. m
M60.235	Wood partly glazed doors or screens						
	first coat to untreated surfaces; over 300 mm girth						
	small panes (not exceeding 0.1 m^2)	0.30	—	—	—	0.065	sq. m
	medium panes (0.1–0.5 m^2)	0.22	—	—	—	0.054	sq. m
	large panes (0.5–1.0 m^2)	0.20	—	—	—	0.043	sq. m
	extra large panes (over 1.0 m^2)	0.19	—	—	—	0.038	sq. m
	second and subsequent coats; over 300 mm girth						
	small panes (not exceeding 0.1 m^2)	0.30	—	—	—	0.051	sq. m
	medium panes (0.1–0.5 m^2)	0.22	—	—	—	0.043	sq. m
	large panes (0.5–1.0 m^2)	0.20	—	—	—	0.034	sq. m
	extra large panes (over 1.0 m^2)	0.19	—	—	—	0.030	sq. m
M60.240	Wood windows						
	first coat to untreated surfaces; over 300 mm girth						
	small panes (not exceeding 0.1 m^2)	0.33	—	—	—	0.086	sq. m
	medium panes (0.1–0.5 m^2)	0.23	—	—	—	0.076	sq. m
	large panes (0.5–1.0 m^2)	0.21	—	—	—	0.065	sq. m
	extra large panes (over 1.0 m^2)	0.20	—	—	—	0.054	sq. m
	edges of opening casements	0.04	—	—	—	0.016	ln. m
	second and subsequent coats; over 300 mm girth						
	small panes (not exceeding 0.1 m^2)	0.33	—	—	—	0.068	sq. m
	medium panes (0.1–0.5 m^2)	0.23	—	—	—	0.060	sq. m
	large panes (0.5–1.0 m^2)	0.21	—	—	—	0.051	sq. m
	extra large panes (over 1.0 m^2)	0.20	—	—	—	0.043	sq. m
	edges of opening casements	0.04	—	—	—	0.013	ln. m

N FURNITURE

NA   **Notes**

I   The labour constants in this section are based on one craftsman joiner.

N11 *DOMESTIC KITCHEN FITTINGS*

N11.005   **Kitchen cupboard units, ready assembled**

N11.010   Floor cupboard units, 600 mm deep, 870 mm high

N11.015   Floor cupboard units, 600 mm deep, 870 mm high

N11.020   Tall cupboard units, 600 mm deep, 2100 mm high

N11.025   Wall cupboard units, 300 mm deep, 561 or 724 mm high

N11.030   Corner wall cupboard units, 300 mm deep, 561 or 724 mm high

N11.035   Worktops; 500 or 600 wide

N11.040   Sundry accessories

N11   DOMESTIC KITCHEN FITTINGS

	Labour (hr)	Unit (no.)	Worktop (no.)	Sundries (no.)	Sundries (ln. m)	Unit
N11.005 **Kitchen cupboard units, ready assembled**						
N11.010 Floor cupboard units, 600 mm deep, 870 mm high; base units						
300 mm long	0.67	1.00	—	—	—	no.
400 mm long	0.75	1.00	—	—	—	no.
500 mm long	0.83	1.00	—	—	—	no.
600 mm long	1.08	1.00	—	—	—	no.
800 mm long	1.33	1.00	—	—	—	no.
1000 mm long	1.50	1.00	—	—	—	no.
1200 mm long	1.67	1.00	—	—	—	no.
N11.015 Floor cupboard units, 600 mm deep, 870 mm high						
sink base units						
800 mm long	1.33	1.00	—	—	—	no.
1000 mm long	1.50	1.00	—	—	—	no.
1200 mm long	1.67	1.00	—	—	—	no.
corner base units						
1000 mm long	1.75	1.00	—	—	—	no.
1200 mm long	2.00	1.00	—	—	—	no.
N11.020 Tall cupboard units, 600 mm deep, 2100 mm high						
storage units						
500 mm long	1.69	1.00	—	—	—	no.
600 mm long	1.75	1.00	—	—	—	no.
oven housing units						
600 mm long	1.75	1.00	—	—	—	no.
N11.025 Wall cupboard units, 300 mm deep						
561 mm high						
300 mm long	0.83	1.00	—	—	—	no.
400 mm long	1.00	1.00	—	—	—	no.
500 mm long	1.17	1.00	—	—	—	no.
600 mm long	1.33	1.00	—	—	—	no.
800 mm long	1.67	1.00	—	—	—	no.
1000 mm long	1.94	1.00	—	—	—	no.
724 mm high						
300 mm long	0.91	1.00	—	—	—	no.
400 mm long	1.10	1.00	—	—	—	no.
500 mm long	1.29	1.00	—	—	—	no.
600 mm long	1.46	1.00	—	—	—	no.
800 mm long	1.84	1.00	—	—	—	no.
1000 mm long	2.20	1.00	—	—	—	no.

	Labour (hr)	Unit (no.)	Worktop (ln. m)	Sundries (no.)	Sundries (ln. m)	Unit
N11.030 Corner wall cupboard units, 300 mm deep						
561 mm high						
600 mm long	2.33	1.00	—	—	—	no.
724 mm high;						
600 mm long	2.56	1.00	—	—	—	no.
N11.035 Worktops; 500 or 600 mm wide						
300 mm long	0.30	—	0.30	—	—	no.
400 mm long	0.32	—	0.40	—	—	no.
500 mm long	0.33	—	0.50	—	—	no.
600 mm long	0.41	—	0.60	—	—	no.
800 mm long	0.51	—	0.80	—	—	no.
1000 mm long	0.58	—	1.00	—	—	no.
1200 mm long	0.73	—	1.20	—	—	no.
N11.040 Sundry accessories						
end panels, 600 mm wide × 900 mm high	0.58	—	—	—	1.00	no.
leg support with No. 2 ferrules	0.33	—	—	—	1.00	no.
tray space plinth, up to 300 mm wide	0.83	—	—	—	1.00	no.
tidy bins, vegetable racks or the like	0.25	—	—	—	1.00	no.
cornices	0.33	—	—	1.05	—	ln. m
lighting pelmets	0.17	—	—	1.05	—	ln. m
mitres on cornices	0.33	—	—	—	—	no.
mitres on lighting pelmets	0.25	—	—	—	—	no.

P BUILDING FABRIC SUNDRIES

PA   **Notes**

I   The labour constants in this section are based on one craftsman joiner.

P10 *SUNDRY INSULATION*

P10.005   **Glass fibre insulation**

P10.010   Fixing between members at 600 mm centres; horizontally

P10.015   Fixing between members at 600 mm centres; vertically

P10.020   **Glass fibre sound deadening quilt type PF**

P10.025   Fixing in position; horizontally

P10.030   Fixing in position; vertically

P10   SUNDRY INSULATION

		Labour (hr)				Insulation (sq. m)	Unit
P10.005	**Glass fibre insulation**						
P10.010	Fixing between members at 600 mm centres; horizontally						
	100 mm thick	0.09	—	—	—	1.05	sq. m
	150 mm thick	0.10	—	—	—	1.05	sq. m
P10.015	Fixing between members at 600 mm centres; vertically						
	100 mm thick	0.09	—	—	—	1.05	sq. m
	150 mm thick	0.10	—	—	—	1.05	sq. m
P10.020	**Glass fibre sound deadening quilt type PF**						
P10.025	Fixing in position; horizontally						
	13 mm thick	0.07	—	—	—	1.05	sq. m
	25 mm thick	0.08	—	—	—	1.05	sq. m
P10.030	Fixing in position; vertically						
	13 mm thick	0.07	—	—	—	1.05	sq. m
	25 mm thick	0.08	—	—	—	1.05	sq. m

P20 *UNFRAMED ISOLATED TRIMS/SKIRTINGS/SUNDRY ITEMS*

P20.005	**Wrought softwood**
P20.010	Skirtings, rails or the like
P20.015	Architraves, cover fillets or the like
P20.020	Bearers
P20.025	Slats
P20.030	Window boards, nosings, bed moulds or the like
P20.035	Notched and returned ends
P20.105	**Blockboard, BS.3444, 2/3 grade, BR bonded**
P20.110	Isolated shelves, worktops, seats or the like
P20.115	Labour on blockboard
P20.205	**Plywood, BS.1455, 2/3 grade, WBP bonded**
P20.210	Isolated shelves, worktops, seats or the like
P20.215	Labour on plywood
P20.305	**Chipboard, BS.5669**
P20.310	Isolated shelves, worktops, seats or the like
P20.315	Labour on chipboard
P20.405	**Melamine faced chipboard**
P20.410	Isolated shelves, worktops, seats or the like
P20.415	Labour on melamine faced chipboard

P20   UNFRAMED ISOLATED TRIMS/SKIRTINGS/SUNDRY ITEMS

	Labour (hr)			Timber (ln. m)	Nails (kg)	Unit
P20.005	**Wrought softwood**					
P20.010	Skirtings, rails or the like					
13 × 50 mm	0.37	—	—	1.075	0.010	ln. m
13 × 75 mm	0.38	—	—	1.075	0.010	ln. m
19 × 50 mm	0.38	—	—	1.075	0.010	ln. m
19 × 75 mm	0.40	—	—	1.075	0.030	ln. m
19 × 100 mm	0.42	—	—	1.075	0.030	ln. m
25 × 50 mm	0.40	—	—	1.075	0.010	ln. m
25 × 75 mm	0.42	—	—	1.075	0.030	ln. m
25 × 100 mm	0.43	—	—	1.075	0.030	ln. m
25 × 125 mm	0.45	—	—	1.075	0.030	ln. m
25 × 150 mm	0.48	—	—	1.075	0.030	ln. m
25 × 175 mm	0.50	—	—	1.075	0.040	ln. m
25 × 200 mm	0.53	—	—	1.075	0.040	ln. m
25 × 225 mm	0.56	—	—	1.075	0.040	ln. m
P20.015	Architraves, cover fillets or the like					
13 × 13 mm	0.12	—	—	1.075	0.010	ln. m
13 × 19 mm	0.13	—	—	1.075	0.010	ln. m
13 × 25 mm	0.13	—	—	1.075	0.010	ln. m
13 × 32 mm	0.14	—	—	1.075	0.010	ln. m
13 × 50 mm	0.14	—	—	1.075	0.010	ln. m
19 × 19 mm	0.15	—	—	1.075	0.010	ln. m
19 × 25 mm	0.15	—	—	1.075	0.010	ln. m
19 × 32 mm	0.16	—	—	1.075	0.010	ln. m
19 × 50 mm	0.17	—	—	1.075	0.010	ln. m
19 × 75 mm	0.18	—	—	1.075	0.010	ln. m
25 × 25 mm	0.15	—	—	1.075	0.010	ln. m
25 × 32 mm	0.16	—	—	1.075	0.010	ln. m
25 × 50 mm	0.17	—	—	1.075	0.010	ln. m
25 × 75 mm	0.18	—	—	1.075	0.010	ln. m
25 × 100 mm	0.19	—	—	1.075	0.010	ln. m
25 × 125 mm	0.20	—	—	1.075	0.010	ln. m
25 × 150 mm	0.21	—	—	1.075	0.010	ln. m

	Labour (hr)			Timber (ln. m)	Nails (kg)	Unit
P20.020 Bearers						
25 × 32 mm	0.13	—	—	1.075	0.010	ln. m
25 × 50 mm	0.14	—	—	1.075	0.010	ln. m
32 × 32 mm	0.15	—	—	1.075	0.010	ln. m
32 × 50 mm	0.17	—	—	1.075	0.010	ln. m
50 × 50 mm	0.18	—	—	1.075	0.010	ln. m
50 × 75 mm	0.20	—	—	1.075	0.010	ln. m
75 × 75 mm	0.22	—	—	1.075	0.010	ln. m
75 × 100 mm	0.25	—	—	1.075	0.010	ln. m
100 × 100 mm	0.29	—	—	1.075	0.010	ln. m
P20.025 Slats						
25 × 38 mm	0.13	—	—	1.075	—	ln. m
25 × 50 mm	0.13	—	—	1.075	—	ln. m
P20.030 Window boards, nosings, bed moulds or the like						
25 × 75 mm	0.57	—		1.075	0.03	ln. m
25 × 100 mm	0.61	—		1.075	0.03	ln. m
25 × 125 mm	0.63	—		1.075	0.03	ln. m
25 × 150 mm	0.67	—		1.075	0.04	ln. m
25 × 175 mm	0.71	—		1.075	0.04	ln. m
25 × 200 mm	0.74	—		1.075	0.05	ln. m
25 × 225 mm	0.77	—		1.075	0.05	ln. m
25 × 250 mm	0.80	—		1.075	0.05	ln. m
32 × 75 mm	0.67	—		1.075	0.03	ln. m
32 × 100 mm	0.71	—		1.075	0.03	ln. m
32 × 125 mm	0.74	—		1.075	0.03	ln. m
32 × 150 mm	0.77	—		1.075	0.04	ln. m
32 × 175 mm	0.83	—		1.075	0.04	ln. m
32 × 200 mm	0.87	—		1.075	0.05	ln. m
32 × 225 mm	0.91	—		1.075	0.05	ln. m
32 × 250 mm	0.95	—		1.075	0.05	ln. m
P20.035 Notched and returned ends	0.25	—	—	—	—	no.

	Labour (hr)			Boarding (sq. m)	Nails (kg)	Unit
P20.105	**Blockboard, BS.3444, 2/2 grade, BR bonded; butt joints**					
P20.110	Isolated shelves, worktops, seats or the like					
	12 mm thick					
	not exceeding 150 mm wide	0.13	—	—	0.17	—	ln. m
	150–300 mm wide	0.26	—	—	0.35	—	ln. m
	over 300 mm wide	0.69	—	—	1.15	—	sq. m
	18 mm thick					
	not exceeding 150 mm wide	0.17	—	—	0.17	—	ln. m
	150–300 mm wide	0.34	—	—	0.35	—	ln. m
	over 300 mm wide	0.92	—	—	1.15	—	sq. m
	25 mm thick					
	not exceeding 150 mm wide	0.22	—	—	0.17	—	ln. m
	150–300 mm wide	0.43	—	—	0.35	—	ln. m
	over 300 mm wide	1.15	—	—	1.15	—	sq. m
P20.115	Labour on blockboard					
	12 mm thick					
	raking cutting	0.32	—	—	—	—	ln. m
	curved cutting	0.64	—	—	—	—	ln. m
	notches, per 25 mm girth	0.21	—	—	—	—	no.
	forming openings, not exceeding 0.50 m^2	0.42	—	—	—	—	no.
	18 mm thick					
	raking cutting	0.32	—	—	—	—	ln. m
	curved cutting	0.64	—	—	—	—	ln. m
	notches, per 25 mm girth	0.21	—	—	—	—	no.
	forming openings, not exceeding 0.50 m^2	0.42	—	—	—	—	no.
	25 mm thick					
	raking cutting	0.35	—	—	—	—	ln. m
	curved cutting	0.70	—	—	—	—	ln. m
	notches, per 25 mm girth	0.23	—	—	—	—	no.
	forming openings, not exceeding 0.50 m^2	0.46	—	—	—	—	no.

	Labour (hr)			Boarding (sq. m)	Nails (kg)	Unit
P20.205	**Plywood, BS.1455, 2/3 grade, WBP bonded; butt joints**					
P20.210	Isolated shelves, worktops, seats or the like					
	6 mm thick					
	not exceeding 150 mm wide	0.10	—	—	0.17	—	ln. m
	150–300 mm wide	0.19	—	—	0.35	—	ln. m
	over 300 mm wide	0.50	—	—	1.15	—	sq. m
	9 mm thick					
	not exceeding 150 mm wide	0.12	—	—	0.17	—	ln. m
	150–300 mm wide	0.24	—	—	0.35	—	ln. m
	over 300 mm wide	0.63	—	—	1.15	—	sq. m
	12 mm thick					
	not exceeding 150 mm wide	0.14	—	—	0.17	—	ln. m
	150–300 mm wide	0.29	—	—	0.35	—	ln. m
	over 300 mm wide	0.74	—	—	1.15	—	sq. m
P20.215	Labour on plywood					
	4 mm thick					
	raking cutting	0.25	—	—	—	—	ln. m
	curved cutting	0.50	—	—	—	—	ln. m
	notches, per 25 mm girth	0.17	—	—	—	—	no.
	forming openings, not exceeding 0.50 m^2	0.34	—	—	—	—	no.
	6 mm thick					
	raking cutting	0.25	—	—	—	—	ln. m
	curved cutting	0.50	—	—	—	—	ln. m
	notches, per 25 mm girth	0.17	—	—	—	—	no.
	forming openings, not exceeding 0.50 m^2	0.34	—	—	—	—	no.
	9 mm thick					
	raking cutting	0.29	—	—	—	—	ln. m
	curved cutting	0.58	—	—	—	—	ln. m
	notches, per 25 mm girth	0.19	—	—	—	—	no.
	forming openings, not exceeding 0.50 m^2	0.38	—	—	—	—	no.
	12 mm thick					
	raking cutting	0.32	—	—	—	—	ln. m
	curved cutting	0.64	—	—	—	—	ln. m
	notches, per 25 mm girth	0.21	—	—	—	—	no.
	forming openings, not exceeding 0.50 m^2	0.42	—	—	—	—	no.

	Labour (hr)			Boarding (sq. m)	Nails (kg)	Unit
P20.305	**Chipboard, BS.5669; butt joints**					
P20.310	Isolated shelves, worktops, seats or the like					
	12 mm thick					
	not exceeding 150 mm wide	0.13	—	—	0.17	—	ln. m
	150–300 mm wide	0.26	—	—	0.35	—	ln. m
	over 300 mm wide	0.69	—	—	1.15	—	sq. m
	18 mm thick					
	not exceeding 150 mm wide	0.17	—	—	0.17	—	ln. m
	150–300 mm wide	0.34	—	—	0.35	—	ln. m
	over 300 mm wide	0.92	—	—	1.15	—	sq. m
P20.315	Labour on chipboard					
	12 mm thick					
	raking cutting	0.29	—	—	—	—	ln. m
	curved cutting	0.58	—	—	—	—	ln. m
	notches, per 25 mm girth	0.19	—	—	—	—	no.
	forming openings, not exceeding 0.50 m^2	0.38	—	—	—	—	no.
	18 mm thick					
	raking cutting	0.29	—	—	—	—	ln. m
	curved cutting	0.58	—	—	—	—	ln. m
	notches, per 25 mm girth	0.19	—	—	—	—	no.
	forming openings, not exceeding 0.50 m^2	0.38	—	—	—	—	no.

		Labour (hr)			Boarding (sq. m)	Nails (kg)	Unit
P20.405	**Melamine faced chipboard; butt joints**						
P20.410	Isolated shelves, worktops, seats or the like						
	15 mm thick						
	not exceeding 150 mm wide	0.16	—	—	0.17	—	ln. m
	150–300 mm wide	0.33	—	—	0.35	—	ln. m
	over 300 mm wide	0.87	—	—	1.15	—	sq. m
	18 mm thick						
	not exceeding 150 mm wide	0.22	—	—	0.17	—	ln. m
	150–300 mm wide	0.44	—	—	0.35	—	ln. m
	over 300 mm wide	1.16	—	—	1.15	—	sq. m
P20.415	Labour on melamine faced chipboard						
	15 mm thick						
	raking cutting	0.29	—	—	—	—	ln. m
	curved cutting	0.58	—	—	—	—	ln. m
	notches, per 25 mm girth	0.19	—	—	—	—	no.
	forming openings, not exceeding 0.50 m^2	0.38	—	—	—	—	no.
	18 mm thick						
	raking cutting	0.29	—	—	—	—	ln. m
	curved cutting	0.58	—	—	—	—	ln. m
	notches, per 25 mm girth	0.19	—	—	—	—	no.
	forming openings, not exceeding 0.50 m^2	0.38	—	—	—	—	no.

P21   *IRONMONGERY*

P21 IRONMONGERY

	Labour (hr)			Ironmongery (no.)	Screws (no.)	Unit
P21.005 **Ironmongery; to softwood**						
P21.010 Hinges						
steel butts						
75 mm	0.20	—	—	1.025	13.20	pair
100 mm	0.22	—	—	1.025	17.60	pair
cast iron butts						
75 mm	0.22	—	—	1.025	13.20	pair
100 mm	0.24	—	—	1.025	17.60	pair
parliament hinges						
75 mm	0.33	—	—	1.025	13.20	pair
100 mm	0.37	—	—	1.025	17.60	pair
steel rising butts						
75 mm	1.33	—	—	1.025	13.20	pair
100 mm	1.42	—	—	1.025	17.60	pair
backflap hinges						
38 mm	1.00	—	—	1.025	13.20	pair
50 mm	1.00	—	—	1.025	13.20	pair
tee hinges, medium						
300 mm	0.58	—	—	1.025	15.40	pair
450 mm	0.83	—	—	1.025	19.80	pair
600 mm	0.87	—	—	1.025	19.80	pair
hooks and bands						
450 mm	1.25	—	—	1.025	19.80	pair
600 mm	1.33	—	—	1.025	19.80	pair
P21.015 Floor springs; setting into prepared mortice						
single action	2.00	—	—	1.025	8.80	no.
double action	2.33	—	—	1.025	8.80	no.
P21.020 Locks and latches						
rim latch and backset	1.17	—	—	1.025	8.80	no.
rim dead lock and backset	0.86	—	—	1.025	8.80	no.
mortice latch, tubular	0.86	—	—	1.025	4.40	no.
mortice latch, rebated	0.98	—	—	1.025	4.40	no.
mortice lock, upright	1.03	—	—	1.025	4.40	no.
mortice lock, upright, rebated	1.50	—	—	1.025	4.40	no.
cylinder rim night latch	1.73	—	—	1.025	4.40	no.
ball catch	0.35	—	—	1.025	4.40	no.
bales catch	0.57	—	—	1.025	4.40	no.
mortice cupboard lock	0.57	—	—	1.025	4.40	no.
mortice cupboard lock, rebated	0.81	—	—	1.025	4.40	no.
suffolk latch, light	1.15	—	—	1.025	13.20	no.
gate latch	0.29	—	—	1.025	13.20	no.

		Labour (hr)			Iron-mongery (no.)	Screws (no.)	Unit
P21.025	Door furniture						
	lever latch furniture	0.46	—	—	1.025	8.80	set
	lever lock furniture	0.46	—	—	1.025	8.80	set
	escutcheons	0.17	—	—	1.025	2.20	no.
P21.030	Door closers						
	overhead, single action	1.26	—	—	1.025	6.60	no.
	overhead, double action	1.38	—	—	1.025	6.60	no.
	door spring, 203 mm	0.38	—	—	1.025	6.60	no.
	'Perko' concealed closer	0.40	—	—	1.025	8.80	no.
P21.035	Bolts						
	barrel or tower bolts						
	100 mm	0.23	—	—	1.025	8.80	no.
	150 mm	0.29	—	—	1.025	8.80	no.
	200 mm	0.35	—	—	1.025	8.80	no.
	225 mm	0.40	—	—	1.025	8.80	no.
	250 mm	0.46	—	—	1.025	8.80	no.
	monkey tail bolts						
	450 mm	1.15	—	—	1.025	11.00	no.
	600 mm	1.44	—	—	1.025	11.00	no.
	flush bolts						
	100 mm	0.69	—	—	1.025	11.00	no.
	150 mm	0.92	—	—	1.025	11.00	no.
	200 mm	1.15	—	—	1.025	11.00	no.
	225 mm	1.38	—	—	1.025	11.00	no.
	250 mm	1.61	—	—	1.025	11.00	no.
	WC indicator bolt	1.15	—	—	1.025	6.60	no.
	panic bolt, single door	2.07	—	—	1.025	18.70	no.
	panic bolt, double door	3.22	—	—	1.025	26.40	no.
	easy-clean bolt socket (mortice measured separately)	0.23	—	—	1.025	2.20	no.
P21.040	Handles and pulls						
	cupboard knobs	0.14	—	—	1.025	—	no.
	pull handles						
	150 mm	0.29	—	—	1.025	6.60	no.
	225 mm	0.35	—	—	1.025	6.60	no.
	300 mm	0.38	—	—	1.025	6.60	no.
	450 mm	0.40	—	—	1.025	6.60	no.
	600 mm	0.46	—	—	1.025	6.60	no.

	Labour (hr)			Iron-mongery (no.)	Screws (no.)	Unit
P21.045 Plates						
push plates						
200 mm	0.23	—	—	1.025	4.40	no.
300 mm	0.23	—	—	1.025	4.40	no.
600 mm	0.29	—	—	1.025	6.60	no.
kicking plates						
450 × 225 mm	0.40	—	—	1.025	6.60	no.
600 × 225 mm	0.44	—	—	1.025	6.60	no.
750 × 225 mm	0.46	—	—	1.025	8.80	no.
900 × 225 mm	0.48	—	—	1.025	10.10	no.
P21.050 Letterplate						
including cutting opening	1.44	—	—	1.025	4.40	no.
P21.055 Window fittings						
casement stay	0.29	—	—	1.025	2.20	no.
casement fastener	0.29	—	—	1.025	4.40	no.
quadrant stays	0.35	—	—	1.025	3.30	no.
fanlight catch	0.29	—	—	1.025	4.40	no.
sash fastener	0.29	—	—	1.025	4.40	no.
sash lift	0.17	—	—	1.025	3.30	no.
sash pulley	0.35	—	—	1.025	2.20	no.
P21.060 Sundry items						
hat and coat hooks	0.17	—	—	1.025	2.20	no.
cabin hook and eye	0.23	—	—	1.025	8.80	no.
hasps and staples	0.35	—	—	1.025	8.80	no.
rubber door stops	0.12	—	—	1.025	1.10	no.
shelf brackets	0.35	—	—	1.025	6.60	no.
curtain track	0.77	—	—	1.025	—	ln. m
draught excluder	0.38	—	—	1.025	—	ln. m
hanging rail	0.12	—	—	1.025	—	ln. m
sockets	0.17	—	—	1.025	3.30	no.
centre brackets	0.23	—	—	1.025	2.20	no.

Q PAVING

QA

Notes

I

The labour constants in this section are based on one labourer.

Q10   *CONCRETE KERBS/EDGINGS/CHANNELS*

Q10.005   **Precast concrete; standard or stock pattern units; BS.340**

Q10.010   Kerbs; 125 × 255 mm; Figures 2, 5, and 7

Q10.015   Kerbs; 125 × 150 mm; Figures 2a, 7a, 8 and 9

Q10.020   Kerbs; tapering; 915 mm; 125 × 255 to 150 mm; Figure 16

Q10.025   **Precast concrete; standard or stock pattern units; BS.340**

Q10.030   Edgings; 50 × 150 mm; Figures 11, 12 and 13

Q10.105   **Precast concrete; standard or stock pattern units; BS.340**

Q10.110   Channels; 125 × 255 mm; Figure 8

Q10.115   Channels; 125 × 150 mm; Figure 8

Q10   CONCRETE/KERBS/EDGINGS/CHANNELS

		Labour (hr)		Kerbs (no.)	Cement mortar (cu. m)	Concrete (cu. m)	Unit
Q10.005	**Precast concrete; standard or stock pattern units; BS.340; bedding, jointing and pointing in cement mortar; haunching with concrete one side; formwork**						
Q10.010	Kerbs; 125 × 255 mm; Figures 2, 5 and 7						
	straight	0.55	—	1.120	0.003	0.029	ln. m
	curved; radius						
	not exceeding 3.00 m	0.78	—	1.120	0.003	0.029	ln. m
	3.00–6.00 m	0.73	—	1.120	0.003	0.029	ln. m
	6.00–9.00 m	0.68	—	1.120	0.003	0.029	ln. m
	over 9.00 m	0.61	—	1.120	0.003	0.029	ln. m
Q10.015	Kerbs; 125 × 150 mm; Figures 2a, 7a, 8 and 9						
	straight	0.53	—	1.120	0.003	0.014	ln. m
	curved; radius						
	not exceeding 3.00 m	0.74	—	1.120	0.003	0.014	ln. m
	3.00–6.00 m	0.69	—	1.120	0.003	0.014	ln. m
	6.00–9.00 m	0.64	—	1.120	0.003	0.014	ln. m
	over 9.00 m	0.58	—	1.120	0.003	0.014	ln. m
Q10.020	Kerbs; tapering; 915 mm long						
	125 × 255 to 150 mm; Figure 16	0.53	—	1.025	0.003	0.013	no.
Q10.025	**Precast concrete; standard or stock pattern units; BS.340; bedding, jointing and pointing in cement mortar; haunching with concrete both sides; formwork**						
Q10.030	Edgings; 50 × 150 mm; Figures 11, 12 and 13						
	straight	0.40	—	1.120	0.001	0.013	ln. m
	curved; radius						
	not exceeding 3.00 m	0.56	—	1.120	0.001	0.013	ln. m
	3.00–6.00 m	0.52	—	1.120	0.001	0.013	ln. m
	6.00–9.00 m	0.48	—	1.120	0.001	0.013	ln. m
	over 9.00 m	0.44	—	1.120	0.001	0.013	ln. m

		Labour (hr)		Kerbs (no.)	Cement mortar (cu. m)	Concrete (cu. m)	Unit
Q10.105	**Precast concrete; standard or stock pattern units; BS.340; bedding, jointing and pointing in cement mortar; haunching with concrete one side; formwork**						
Q10.110	Channels; 125 × 255 mm; Figure 8						
	straight	0.50	—	1.120	0.006	0.003	ln. m
	curved; radius						
	not exceeding 3.00 m	0.71	—	1.120	0.006	0.003	ln. m
	3.00–6.00 m	0.67	—	1.120	0.006	0.003	ln. m
	6.00–9.00 m	0.63	—	1.120	0.006	0.003	ln. m
	over 9.00 m	0.56	—	1.120	0.006	0.003	ln. m
Q10.115	Channels; 125 × 150 mm; Figure 8						
	straight	0.45	—	1.120	0.003	0.004	ln. m
	curved; radius						
	not exceeding 3.00 m	0.65	—	1.120	0.003	0.004	ln. m
	3.00–6.00 m	0.60	—	1.120	0.003	0.004	ln. m
	6.00–9.00 m	0.56	—	1.120	0.003	0.004	ln. m
	over 9.00 m	0.50	—	1.120	0.003	0.004	ln. m

Q20   *HARDCORE/GRANULAR/SUB-BASES TO ROADS/PAVINGS*

Q20.005	**Filling with material arising from excavations**
Q20.010	Filling to excavations, by hand, compacting in layers
Q20.015	Filling to make up levels, by hand, compacting in layers
Q20.020	Filling to make up levels, by machine (JCB), compacting in layers
Q20.025	**Imported sand filling**
Q20.030	Filling to excavations, by hand, compacting in layers
Q20.035	Filling to make up levels, by hand, compacting in layers
Q20.040	Filling to make up levels, by machine (JCB), compacting in layers
Q20.050	**Imported hardcore filling**
Q20.055	Filling to make up levels, by hand, compacting in layers
Q20.060	Filling to make up levels, by machine (JCB), compacting in layers
Q20.070	**Imported granular fill material, M.O.T. Type I**
Q20.075	Filling to make up levels, by machine (JCB), compacting with a 6.00–8.00 tonne roller

Q20.105	**Surface treatments**
Q20.110	Compacting ground with a whacker
Q20.115	Compacting ground with a vibrating roller
Q20.120	Compacting filling with a whacker
Q20.125	Compacting filling with a vibrating roller
Q20.130	Compacting filling with a 6.00–8.00 tonne roller
Q20.135	Compacting bottoms of excavations with a whacker
Q20.140	Compacting bottoms of excavations with a vibrating roller
Q20.145	Blinding surfaces of filling with sand and consolidating
Q20.150	Hand packing hardcore to form vertical or battering faces

Q20   HARDCORE/GRANULAR/SUB-BASES TO ROADS/PAVINGS

	Labour (hr)	Machine excavating (hr)	Plant compacting (hr)	Filling sand (tonne)	Unit
Q20.005 **Filling with material arising from excavations**					
Q20.010 Filling to excavations, by hand, compacting with whacker in 250 mm layers	1.25	—	0.20	—	—	cu. m
Q20.015 Filling to make up levels by hand, wheeling average 25 m, compacting with a vibrating roller					
average 75 mm thick	0.20	—	0.05	—	—	sq. m
average 100 mm thick	0.25	—	0.05	—	—	sq. m
average 125 mm thick	0.30	—	0.05	—	—	sq. m
average 150 mm thick	0.34	—	0.05	—	—	sq. m
average 175 mm thick	0.40	—	0.05	—	—	sq. m
average 200 mm thick	0.43	—	0.05	—	—	sq. m
average 225 mm thick	0.50	—	0.05	—	—	sq. m
average 250 mm thick	0.55	—	0.05	—	—	sq. m
over 250 mm thick	2.00	—	0.20	—	—	cu. m
Q20.020 Filling to make up levels by machine (JCB) transporting average 25 m, compacting with a vibrating roller					
average 75 mm thick	0.07	0.01	0.05	—	—	sq. m
average 100 mm thick	0.07	0.01	0.05	—	—	sq. m
average 125 mm thick	0.08	0.01	0.05	—	—	sq. m
average 150 mm thick	0.08	0.02	0.05	—	—	sq. m
average 175 mm thick	0.09	0.02	0.05	—	—	sq. m
average 200 mm thick	0.09	0.02	0.05	—	—	sq. m
average 225 mm thick	0.10	0.03	0.05	—	—	sq. m
average 250 mm thick	0.10	0.03	0.05	—	—	sq. m
over 250 mm thick	0.20	0.10	0.20	—	—	cu. m

	Labour (hr)	Machine excavating (hr)	Plant compacting (hr)		Filling sand (tonne)	Unit
Q20.025	**Imported sand filling**					
Q20.030	Filling to excavations, by hand, compacting with whacker in 250 mm layers					
	1.00	—	0.20	—	1.936	cu. m
Q20.035	Filling to make up levels by hand, compacting with a vibrating roller					
average 75 mm thick	0.19	—	0.05	—	0.145	sq. m
average 100 mm thick	0.23	—	0.05	—	0.194	sq. m
average 125 mm thick	0.28	—	0.05	—	0.242	sq. m
average 150 mm thick	0.32	—	0.05	—	0.290	sq. m
average 175 mm thick	0.37	—	0.05	—	0.339	sq. m
average 200 mm thick	0.41	—	0.05	—	0.387	sq. m
average 225 mm thick	0.46	—	0.05	—	0.436	sq. m
average 250 mm thick	0.50	—	0.05	—	0.484	sq. m
over 250 mm thick	1.80	—	0.20	—	1.936	cu. m
Q20.040	Filling to make up levels by machine (JCB) compacting with a vibrating roller					
average 75 mm thick	0.07	0.01	0.05	—	0.145	sq. m
average 100 mm thick	0.07	0.01	0.05	—	0.194	sq. m
average 125 mm thick	0.08	0.01	0.05	—	0.242	sq. m
average 150 mm thick	0.08	0.02	0.05	—	0.290	sq. m
average 175 mm thick	0.09	0.02	0.05	—	0.339	sq. m
average 200 mm thick	0.09	0.02	0.05	—	0.387	sq. m
average 225 mm thick	0.10	0.03	0.05	—	0.436	sq. m
average 250 mm thick	0.10	0.03	0.05	—	0.484	sq. m
over 250 mm thick	0.20	0.10	0.20	—	1.936	cu. m

		Labour (hr)	Machine excavating (hr)	Plant compacting (hr)		Hardcore (tonne)	Unit
Q20.050	**Imported hardcore filling**						
Q20.055	Filling to excavations, by hand, compacting with a whacker in 250 mm layers	1.95	—	0.20	—	2.025	cu. m
Q20.060	Filling to make up levels by hand, compacting with a vibrating roller;						
	average 75 mm thick	0.25	—	0.05	—	0.152	sq. m
	average 100 mm thick	0.32	—	0.05	—	0.203	sq. m
	average 125 mm thick	0.38	—	0.05	—	0.253	sq. m
	average 150 mm thick	0.45	—	0.05	—	0.304	sq. m
	average 175 mm thick	0.51	—	0.05	—	0.354	sq. m
	average 200 mm thick	0.58	—	0.05	—	0.405	sq. m
	average 225 mm thick	0.65	—	0.05	—	0.456	sq. m
	average 250 mm thick	0.71	—	0.05	—	0.506	sq. m
	over 250 mm thick	2.65	—	0.20	—	2.025	cu. m
Q20.065	Filling to make up levels by machine (JCB) compacting with a vibrating roller						
	average 75 mm thick	0.08	0.02	0.05	—	0.152	sq. m
	average 100 mm thick	0.09	0.02	0.05	—	0.203	sq. m
	average 125 mm thick	0.10	0.03	0.05	—	0.253	sq. m
	average 150 mm thick	0.11	0.03	0.05	—	0.304	sq. m
	average 175 mm thick	0.11	0.04	0.05	—	0.354	sq. m
	average 200 mm thick	0.13	0.04	0.05	—	0.405	sq. m
	average 225 mm thick	0.14	0.05	0.05	—	0.456	sq. m
	average 250 mm thick	0.15	0.05	0.05	—	0.506	sq. m
	over 250 mm thick	0.40	0.20	0.20	—	2.025	cu. m
Q20.070	**Imported granular fill material, M.O.T. Type 1**						
Q20.075	Filling to make up levels, by machine (JCB) compacting with a 6.00–8.00 tonne roller (in road construction or the like)						
	average 75 mm thick	0.03	0.01	0.01	—	0.152	sq. m
	average 100 mm thick	0.03	0.01	0.01	—	0.203	sq. m
	average 125 mm thick	0.04	0.01	0.01	—	0.253	sq. m
	average 150 mm thick	0.04	0.02	0.01	—	0.304	sq. m
	average 175 mm thick	0.05	0.02	0.02	—	0.354	sq. m
	average 200 mm thick	0.05	0.02	0.02	—	0.405	sq. m
	average 225 mm thick	0.06	0.02	0.02	—	0.456	sq. m
	average 250 mm thick	0.07	0.03	0.03	—	0.506	sq. m
	over 250 mm thick	0.20	0.10	0.05	—	2.025	cu. m

		Labour (hr)		Plant compacting (hr)		Filling sand (tonne)	Unit
Q20.105	**Surface treatments**						
Q20.110	Compacting ground, with a whacker	0.07	—	0.07	—	—	sq. m
Q20.115	Compacting ground, with a vibrating roller	0.08	—	0.08	—	—	sq. m
Q20.120	Compacting filling, with a whacker	0.09	—	0.09	—	—	sq. m
	including grading to falls	0.11	—	0.10	—	—	sq. m
	including grading to falls and crossfalls	0.12	—	0.12	—	—	sq. m
Q20.125	Compacting filling, with a vibrating roller	0.13	—	0.05	—	—	sq. m
	including grading to falls	0.16	—	0.05	—	—	sq. m
	including grading to falls and crossfalls	0.18	—	0.05	—	—	sq. m
Q20.130	Compacting filling, with a 6.00–8.00 tonne roller (in road construction)	0.03	—	0.01	—	—	sq. m
	including grading to falls	0.04	—	0.01	—	—	sq. m
	including grading to falls and crossfalls	0.05	—	0.01	—	—	sq. m
Q20.135	Compacting bottoms of excavations, with a whacker	0.09	—	0.09	—	—	sq. m
Q20.140	Compacting bottoms of excavations, with a vibrating roller	0.07	—	0.05	—	—	sq. m
Q20.145	Blinding surfaces of filling with sand and consolidating with a vibrating roller						
	25 mm thick	0.09	—	0.03	—	0.048	sq. m
	50 mm thick	0.13	—	0.03	—	0.096	sq. m
Q20.150	Hand packing hardcore to form vertical or battering faces						
	average 75 mm high	0.08	—	—	—	—	ln. m
	average 100 mm high	0.10	—	—	—	—	ln. m
	average 125 mm high	0.13	—	—	—	—	ln. m
	average 150 mm high	0.15	—	—	—	—	ln. m
	average 175 mm high	0.18	—	—	—	—	ln. m
	average 200 mm high	0.20	—	—	—	—	ln. m
	average 225 mm high	0.23	—	—	—	—	ln. m
	average 250 mm high	0.25	—	—	—	—	ln. m
	over 250 mm high	0.60	—	—	—	—	sq. m

Q24   *INTERLOCKING BLOCK ROADS/PAVINGS*

Q24.005   **Precast concrete block paving; 200 × 100 mm laid on 50 mm bed of sand**

Q24.010   Laid to falls, cross falls and slopes not exceeding 15 degrees from horizontal

Q24.015   Laid to slopes exceeding 15 degrees from horizontal

Q24.020   Raking cutting

Q24.025   Curved cutting

Q24.030   Perimeter cutting of 45 degrees herrring-bone pattern paving

Q24.035   Cutting and making good around manhole covers, gratings or the like

Q24.105   **BDC concrete kerb-setts bedding in cement mortar; haunching with concrete**

Q24.110   190 × 160 standard kerb-setts, laid to low position

Q24.115   190 × 160 standard kerb-setts, laid to high position

Q24.120   190 × 160 crossover kerb-setts, laid to low position

Q24   INTERLOCKING BLOCK ROADS/PAVINGS

	Labour (hr)	Compacting (hr)	Blocks (no.)	Zone 2 sand (tonne)	Silver sand (tonne)	Unit
Q24.005 **Precast concrete block paving; rectangular, chamfered blocks 200 × 100 mm laid on 50 mm thick screeded bed of sand; compacting with hand-guided vibrating plate compactor**						
Q24.010 Laid to falls, cross falls and slopes not exceeding 15 degrees from horizontal; over 300 mm wide; laid flat in						
half bond						
65 mm thick	1.20	0.05	51.25	0.094	0.005	sq. m
80 mm thick	1.38	0.05	51.25	0.094	0.005	sq. m
parquet						
65 mm thick	1.31	0.05	51.25	0.094	0.005	sq. m
80 mm thick	1.48	0.05	51.25	0.094	0.005	sq. m
90 degrees herring-bone						
65 mm thick	1.25	0.05	51.25	0.094	0.005	sq. m
80 mm thick	1.44	0.05	51.25	0.094	0.005	sq. m
45 degrees herring-bone						
65 mm thick	1.31	0.05	51.25	0.094	0.005	sq. m
80 mm thick	1.48	0.05	51.25	0.094	0.005	sq. m
Q24.015 Laid to slopes exceeding 15 degrees from horizontal; over 300 mm wide; laid flat in						
half bond						
65 mm thick	1.31	0.05	51.25	0.094	0.005	sq. m
80 mm thick	1.48	0.05	51.25	0.094	0.005	sq. m
parquet						
65 mm thick	1.43	0.05	51.25	0.094	0.005	sq. m
80 mm thick	1.65	0.05	51.25	0.094	0.005	sq. m
90 degrees herring-bone						
65 mm thick	1.37	0.05	51.25	0.094	0.005	sq. m
80 mm thick	1.58	0.05	51.25	0.094	0.005	sq. m
45 degrees herring-bone						
65 mm thick	1.43	0.05	51.25	0.094	0.005	sq. m
80 mm thick	1.65	0.05	51.25	0.094	0.005	sq. m
Q24.020 Raking cutting						
65 mm thick	0.08	—	5.13	—	—	ln. m
80 mm thick	0.09	—	5.13	—	—	ln. m

		Labour (hr)	Compacting (hr)	Blocks (no.)	Zone 2 sand (tonne)	Silver sand (tonne)	Unit
Q24.025	Curved cutting						
	65 mm thick	0.10	—	5.13	—	—	ln. m
	80 mm thick	0.12	—	5.13	—	—	ln. m
Q24.030	Perimeter cutting of 45 degrees herring-bone pattern paving						
	65 mm thick	0.08	—	5.13	—	—	ln. m
	80 mm thick	0.09	—	5.13	—	—	ln. m
Q24.035	Cutting and making good around steel columns, angles, pipes, manhole covers, gratings or the like						
	not exceeding 0.30 m girth						
	65 mm thick	0.03	—	1.28	—	—	no.
	80 mm thick	0.04	—	1.28	—	—	no.
	0.30–1.00 m girth						
	65 mm thick	0.06	—	1.79	—	—	no.
	80 mm thick	0.08	—	1.79	—	—	no.
	1.00–2.00 m girth						
	65 mm thick	0.12	—	3.84	—	—	no.
	80 mm thick	0.14	—	3.84	—	—	no.
	over 2.00 m girth						
	65 mm thick	0.08	—	2.56	—	—	ln. m
	80 mm thick	0.09		2.56	—	—	ln. m

		Labour (hr)		Setts (no.)	Mortar (cu. m)	Concrete (cu. m)	Unit
Q24.105	**BDC concrete kerb-setts; bedding in cement mortar 25 mm thick; pointing with cement mortar; haunching with in situ concrete, 20 mm aggregate, one side; formwork**						
Q24.110	190 × 160 standard kerb-setts, laid to low position						
	straight	0.37	—	9.76	0.005	0.018	ln. m
	curved; radius						
	not exceeding 3.00 m	0.55	—	9.76	0.005	0.018	ln. m
	3.00–6.00 m	0.49	—	9.76	0.005	0.018	ln. m
	6.00–9.00 m	0.46	—	9.76	0.005	0.018	ln. m
	over 9.00 m	0.44	—	9.76	0.005	0.018	ln. m
	angles						
	Type EC/L	0.04	—	1.03	0.001	—	no.
	Type IC/L	0.08	—	1.03	0.001	0.008	no.
Q24.115	190 × 160 standard kerb-setts, laid to high position						
	straight	0.40	—	9.76	0.004	0.023	ln. m
	curved; radius						
	not exceeding 3.00 m	0.58	—	9.76	0.004	0.023	ln. m
	3.00–6.00 m	0.52	—	9.76	0.004	0.023	ln. m
	6.00–9.00 m	0.49	—	9.76	0.004	0.023	ln. m
	over 9.00 m	0.47	—	9.76	0.004	0.023	ln. m
	angles						
	Type EC/H	0.04	—	1.03	0.001	—	no.
	Type IC/H	0.08	—	1.03	0.001	0.010	no.
Q24.120	190 × 160 cossover kerb-setts, laid to low position						
	straight	0.37	—	9.76	0.005	0.018	ln. m
	curved; radius						
	not exceeding 3.00 m	0.55	—	9.76	0.005	0.018	ln. m
	3.00–6.00 m	0.49	—	9.76	0.005	0.018	ln. m
	6.00–9.00 m	0.46	—	9.76	0.005	0.018	ln. m
	over 9.00 m	0.44	—	9.76	0.005	0.018	ln. m
	angles						
	Type EC/X	0.04	—	1.03	0.001	—	no.
	Type IC/X	0.08	—	1.03	0.001	0.008	no.

Q25 *SLAB/PAVINGS*

Q25.005   **Precast concrete flags to BS.368; 50 mm thick; spot bedded in cement mortar**

Q25.010   10 mm joints, pointing with cement mortar

Q25.015   Close butt joints

Q25.020   Raking cutting

Q25.025   Curved cutting

Q25.030   Cutting and making good around manhole covers, gratings or the like

Q25   SLAB/PAVINGS

		Labour (hr)			Flags (no.)	Cement mortar (cu. m)	Unit
Q25.005	**Precast concrete flags to BS.368 natural finish; 50 mm thick; spot bedding in cement mortar to symmetrical layout; to falls, crossfalls and slopes, not exceeding 15 degrees from the horizontal over 300 mm wide; laid with**						
Q25.010	10 mm joints, pointing with cement mortar						
	450 × 450 mm	1.00	—	—	4.844	0.009	sq. m
	600 × 450 mm	0.67	—	—	3.653	0.007	sq. m
	600 × 600 mm	0.50	—	—	2.755	0.005	sq. m
	600 × 750 mm	0.40	—	—	2.211	0.004	sq. m
	600 × 900 mm	0.33	—	—	1.847	0.004	sq. m
Q25.015	Close butt joints						
	450 × 450 mm	0.83	—	—	5.062	0.005	sq. m
	600 × 450 mm	0.56	—	—	3.796	0.004	sq. m
	600 × 600 mm	0.42	—	—	2.847	0.003	sq. m
	600 × 750 mm	0.33	—	—	2.278	0.002	sq. m
	600 × 900 mm	0.28	—	—	1.898	0.002	sq. m
Q25.020	Raking cutting on						
	450 × 450 mm	0.08	—	—	0.126	—	ln. m
	600 × 450 mm	0.08	—	—	0.095	—	ln. m
	600 × 600 mm	0.08	—	—	0.071	—	ln. m
	600 × 750 mm	0.08	—	—	0.057	—	ln. m
	600 × 900 mm	0.08	—	—	0.047	—	ln. m
Q25.025	Curved cutting on						
	450 × 450 mm	0.10	—	—	0.253	—	ln. m
	600 × 450 mm	0.10	—	—	0.190	—	ln. m
	600 × 600 mm	0.10	—	—	0.142	—	ln. m
	600 × 750 mm	0.10	—	—	0.114	—	ln. m
	600 × 900 mm	0.10	—	—	0.095	—	ln. m
Q25.030	Cutting and making good around steel joists, angles, pipes, tubes, manhole covers, gulley gratings, bollards or the like						
	over 2.0 m girth						
	450 × 450 mm	0.08	—	—	0.345	—	ln. m
	600 × 450 mm	0.08	—	—	0.259	—	ln. m
	600 × 600 mm	0.08	—	—	0.194	—	ln. m
	600 × 750 mm	0.08	—	—	0.162	—	ln. m
	600 × 900 mm	0.08	—	—	0.130	—	ln. m
	not exceeding 0.30 m girth						
	450 × 450 mm	0.03	—	—	0.148	—	no.
	600 × 450 mm	0.03	—	—	0.111	—	no.
	600 × 600 mm	0.03	—	—	0.083	—	no.
	600 × 750 mm	0.03	—	—	0.067	—	no.
	600 × 900 mm	0.03	—	—	0.056	—	no.

	Labour (hr)			Flags (no.)	Cement mortar (cu. m)	Unit
Q25.030 Cutting and making good (continued)						
0.30–1.00 m girth						
450 × 450 mm	0.06	—	—	0.247	—	no.
600 × 450 mm	0.06	—	—	0.185	—	no.
600 × 600 mm	0.06	—	—	0.139	—	no.
600 × 750 mm	0.06	—	—	0.067	—	no.
600 × 900 mm	0.06	—	—	0.056	—	no.
1.00–2.00 m girth						
450 × 450 mm	0.12	—	—	0.494	—	no.
600 × 450 mm	0.12	—	—	0.370	—	no.
600 × 600 mm	0.12	—	—	0.278	—	no.
600 × 750 mm	0.12	—	—	0.222	—	no.
600 × 900 mm	0.12	—	—	0.185	—	no.

R DISPOSAL SYSTEMS

RA	**Notes**

1   The labour constants in Section R10 are based on one craftsman joiner or plumber.

2   The labour constants in Section R12 are based generally on one labourer, with the exception of Sections R12.705–R12.730 which are based on one craftsman bricklayer and one labourer (one and one bricklaying gang).

R10 *RAINWATER PIPEWORK/GUTTERS*

R10.005   **uPVC pipes and fittings; push-fit joints; standard supports**

R10.010   68 mm pipework; joints in running length

R10.015   110 mm pipework; joints in running length

R10.020   Connections to vitrified clay rainwater gulley

R10.025   **uPVC half round gutters and fittings; snap-in joints; standard supports**

R10.030   112 mm gutterwork; joints in running length

R10.035   150 mm gutterwork; joints in running length

R10   RAINWATER PIPEWORK/GUTTERS

	Labour (hr)	Gutter/ pipe (ln. m)	Brackets (no.)	Unions (no.)	Fittings (no.)	Unit
R10.005 **uPVC pipes and fittings; push-fit joints; standard supports**						
R10.010 68 mm pipework; joints in running length						
supports at average 1500 mm centres	0.30	1.05	0.769	0.256	—	ln. m
Extra for shoes	0.10	—	—	—	1.025	no.
Extra for offset bends	0.20	—	—	—	1.025	no.
Extra for branches	0.20	—	—	—	1.025	no.
R10.015 110 mm pipework; joints in running length						
supports at average 1500 mm centres	0.36	1.05	0.769	0.265	—	ln. m
Extra for shoes	0.10	—	—	—	1.025	no.
Extra for offset bends	0.20	—	—	—	1.025	no.
Extra for branches	0.25	—	—	—	1.025	no.
R10.020 Connections to vitrified clay rainwater gulley; joint in cement mortar						
68 mm nominal size pipes	0.20	—	—	—	—	no.
110 mm nominal size pipes	0.20	—	—	—	—	no.
R10.025 **uPVC half round gutters and fittings; snap-in joints; standard supports**						
R10.030 112 mm gutterwork; joints in running length						
supports at average 1000 mm centres	0.30	1.05	1.153	0.256	—	ln. m
Extra for stopped ends	0.10	—	—	—	1.025	no.
Extra for square angles	0.30	—	—	—	1.025	no.
Extra for running outlets	0.30	—	—	—	1.025	no.
R10.035 150 mm gutterwork; joints in running length						
supports at average 1000 mm centres	0.36	1.05	1.153	0.256	—	ln. m
Extra for stopped ends	0.10	—	—	—	1.025	no.
Extra for square angles	0.30	—	—	—	1.025	no.
Extra for running outlets	0.30	—	—	—	1.025	no.

R12 *DRAINAGE BELOW GROUND*

R12.005   **Excavating trenches (by machine) to receive pipes not exceeding 200 mm nominal size**

R12.010   Excavating; filling in above 400 mm thick beds or coverings with excavated material; not exceeding 2.0 m deep

R12.015   Excavating; filling in above 400 mm thick beds or coverings with excavated material; 2.0–4.0 m deep

R12.020   Excavating; filling in above 450 mm thick beds or coverings with excavated material; not exceeding 2.0 m deep

R12.025   Excavating; filling in above 450 mm thick beds or coverings with excavated material; 2.0–4.0 m deep

R12.030   **Excavating trenches (by machine) to receive pipes 225 mm nominal size**

R12.035   Excavating; filling in above 525 mm thick beds or coverings with excavated material; not exceeding 2.0 m deep

R12.040   Excavating; filling in above 525 mm thick beds or coverings with excavated material; 2.0–4.0 m deep

R12.045   **Excavating trenches (by machine) to receive pipes not exceeding 200 mm nominal size**

R12.050   Excavating; filling in above 400 mm thick beds or coverings with granular material; not exceeding 2.0 m deep

R12.055   Excavating; filling in above 400 mm thick beds or coverings with granular material; 2.0–4.0 m deep

R12.060   Excavating; filling in above 450 mm thick beds or coverings with granular material; not exceeding 2.0 m deep

R12.065   Excavating; filling in above 450 mm thick beds or coverings with granular material; 2.0–4.0 m deep

R12.070   **Excavating trenches (by machine) to receive pipes 225 mm nominal size**

R12.075   Excavating; filling in above 525 mm thick beds or coverings with granular material; not exceeding 2.0 m deep

R12.080   Excavating; filling in above 525 mm thick beds or coverings with granular material; 2.0–4.0 m deep

R12.105   **Excavating trenches (by hand) to receive pipes not exceeding 200 mm nominal size**

R12.110   Excavating; filling in above 400 mm thick beds or coverings with excavated material; not exceeding 2.0 m deep

R12.115   Excavating; filling in above 450 mm thick beds or coverings with excavated material; not exceeding 2.0 m deep

R12.120   **Excavating trenches (by hand) to receive pipes 225mm nominal size**

R12.125   Excavating; filling in above 525 mm thick beds or coverings with excavated material; not exceeding 2.0 m deep

R12.130   **Excavating trenches (by hand) to receive pipes not exceeding 200 mm nominal size**

R12.135   Excavating; filling in above 400 mm thick beds or coverings with granular material; not exceeding 2.0 m deep

R12.140   Excavating; filling in above 450 mm thick beds or coverings with granular material; not exceeding 2.0 m deep

R12.145   **Excavating trenches (by hand) to receive pipes 225mm nominal size**

R12.150   Excavating; filling in above 525 mm thick beds or coverings with granular material; not exceeding 2.0 m deep

R12.205	**Beds and coverings; granular material, 10 mm nominal size pea shingle**
R12.210	50 mm bed
R12.215	100 mm bed
R12.220	150 mm bed
R12.225	100 mm bed and filling to half height of pipe
R12.230	150 mm bed and filling to half height of pipe
R12.235	Bed and covering

R12.305	**Beds and coverings; plain in situ concrete**
R12.310	100 mm bed
R12.315	150 mm bed
R12.320	100 mm bed and filling to half height of pipe
R12.325	150 mm bed and filling to half height of pipe
R12.330	Bed and covering

R12.405	**Pipes; vitrified clay pipes and fittings; SuperSleve push-fit polypropylene flexible couplings**
R12.410	100 mm pipework in trenches
R12.415	150 mm pipework in trenches

R12.420	**Vitrified clay accessories**
R12.425	Gullies; HepSleve; joint to pipe
R12.430	Inspection chamber, SuperSleve; joint to pipe

R12.435	**Aluminium alloy accessories**
R12.440	Rodding points, SuperSleve; joint to pipe
R12.445	Gratings; placing in position

R12.450	**Pipes; vitrified clay pipes and fittings; HepSeal; spigot and socket joints with rubber sealing rings**
R12.455	150 mm pipework in trenches
R12.460	225 mm pipework in trenches
R12.465	300 mm pipework in trenches

R12.475	**Concrete accessories**
R12.480	Road gullies; BS.556

R12.485	**Ductile iron accessories**
R12.490	Road gratings and frames; BS.497 grade B

R12.505	**Manholes**
R12.510	Excavating pits (by machine) to receive bases of manholes
R12.515	Excavating pits (by hand) to receive bases of manholes
R12.520	Levelling and compacting bottoms of excavations in manholes
R12.525	Plain in situ concrete beds in manhole bases
R12.530	Plain in situ concrete surrounds to manholes
R12.535	Plain in situ concrete benching in manhole bases
R12.540	Reinforced in situ concrete in suspended slabs to manholes
R12.545	Steel fabric reinforcement in suspended slabs to manholes
R12.550	Circular cutting fabric

R12.605   **Manholes; precast concrete inspection chamber units; Mono**

R12.610   Chamber units

R12.615   Reducer slabs

R12.620   Base shutter units

R12.625   Covers

R12.630   Frames

R12.635   **Manholes; precast concrete manhole chamber units; BS.556**

R12.640   Shaft rings

R12.645   Chamber rings

R12.650   Taper sections

R12.655   Cover slabs

R12.705   **Manholes; engineering bricks**

R12.710   Walls in manholes

R12.715   Extra over engineering bricks for facework

R12.720   Building in ends of pipes

R12.725   **Manholes; mortar, cement and sand**

R12.730   12 mm work to walls on brickwork base; steel trowelled

R12.805	**Manholes; metalwork**
R12.810	Step irons
R12.815	Manhole covers and frames
R12.820	Inspection chamber covers and frames
R12.825	Broadstel universal access covers and frames
R12.905	**Manholes; channels in bottoms; vitrified clay**
R12.910	Half-section, straight
R12.915	Half-section; curved
R12.920	Half-section; channel junction
R12.925	Half-section; tapered
R12.930	Half-section; branch bends
R12.935	Three-quarter-section; branch bends

R12 DRAINAGE BELOW GROUND

	Labour (hr)		Machine excavating (hr)	Plant compacting (hr)	Cart away (cu. m)	Unit

R12.005 **Excavating trenches (by machine) to receive pipes not exceeding 200 mm nominal size; grading bottoms; earthwork support; disposing of surplus excavated material by removing from site**

R12.010 Excavating; filling in above 400 mm thick beds or coverings with material arising from excavations not exceeding 2.0 m deep; average depth

	Labour		Machine excavating	Plant compacting	Cart away	Unit
500 mm	0.08	—	0.08	0.01	0.18	ln. m
750 mm	0.14	—	0.13	0.03	0.18	ln. m
1000 mm	0.20	—	0.18	0.05	0.18	ln. m
1250 mm	0.28	—	0.24	0.08	0.18	ln. m
1500 mm	0.46	—	0.39	0.13	0.24	ln. m
1750 mm	0.54	—	0.46	0.16	0.24	ln. m
2000 mm	0.78	—	0.66	0.24	0.30	ln. m

R12.015 Excavating; filling in above 400 mm thick beds or coverings with material arising from excavations 2.0–4.0 m deep; average depth

	Labour		Machine excavating	Plant compacting	Cart away	Unit
2250 mm	0.95	—	0.81	0.28	0.30	ln. m
2500 mm	1.07	—	0.91	0.32	0.30	ln. m
2750 mm	1.18	—	1.00	0.35	0.30	ln. m
3000 mm	1.29	—	1.10	0.39	0.30	ln. m
3250 mm	1.40	—	1.19	0.43	0.30	ln. m
3500 mm	1.52	—	1.28	0.47	0.30	ln. m
3750 mm	1.63	—	1.38	0.50	0.30	ln. m
4000 mm	1.74	—	1.47	0.54	0.30	ln. m

R12.020 Excavating; filling in above 450 mm thick beds or coverings with material arising from excavations not exceeding 2.0 m deep; average depth

	Labour		Machine excavating	Plant compacting	Cart away	Unit
500 mm	0.08	—	0.08	0.01	0.20	ln. m
750 mm	0.14	—	0.13	0.03	0.20	ln. m
1000 mm	0.20	—	0.17	0.05	0.20	ln. m
1250 mm	0.28	—	0.24	0.07	0.20	ln. m
1500 mm	0.45	—	0.39	0.13	0.27	ln. m
1750 mm	0.53	—	0.46	0.16	0.27	ln. m
2000 mm	0.77	—	0.66	0.23	0.34	ln. m

	Labour (hr)		Machine excavating (hr)	Plant compacting (hr)	Cart away (cu. m)	Unit
R12.025	Excavating; filling in above 450 mm thick beds or coverings with material arising from excavations 2.0–4.0 m deep; average depth					
2250 mm	0.94	—	0.81	0.27	0.34	ln. m
2500 mm	1.06	—	0.90	0.31	0.34	ln. m
2750 mm	1.17	—	1.00	0.34	0.34	ln. m
3000 mm	1.28	—	1.09	0.38	0.34	ln. m
3250 mm	1.39	—	1.19	0.42	0.34	ln. m
3500 mm	1.51	—	1.28	0.46	0.34	ln. m
3750 mm	1.62	—	1.37	0.49	0.34	ln. m
4000 mm	1.73	—	1.47	0.53	0.34	ln. m

	Labour (hr)	Machine excavating (hr)	Plant compacting (hr)	Cart away (cu. m)	Unit
R12.030 **Excavating trenches to receive pipes, 225 mm nominal size; grading bottoms; earthwork support; disposing of surplus excavated material by removing from site**					
R12.035 Excavating; filling in above 525 mm thick beds or coverings with material arising from excavations not exceeding 2.0 m deep; average depth					
750 mm	0.20	—	0.19	0.03	0.32	In. m
1000 mm	0.25	—	0.23	0.06	0.32	In. m
1250 mm	0.36	—	0.31	0.09	0.32	In. m
1500 mm	0.44	—	0.38	0.12	0.32	In. m
1750 mm	0.66	—	0.56	0.18	0.39	In. m
2000 mm	0.76	—	0.65	0.22	0.39	In. m
R12.040 Excavating; filling in above 525 mm thick beds or coverings with material arising from excavations 2.0–4.0 m deep; average depth					
2250 mm	0.93	—	0.81	0.26	0.39	In. m
2500 mm	1.05	—	0.90	0.30	0.39	In. m
2750 mm	1.16	—	0.99	0.33	0.39	In. m
3000 mm	1.27	—	1.09	0.37	0.39	In. m
3250 mm	1.38	—	1.18	0.41	0.39	In. m
3500 mm	1.50	—	1.27	0.45	0.39	In. m
3750 mm	1.61	—	1.37	0.48	0.39	In. m
4000 mm	1.72	—	1.46	0.52	0.39	In. m

		Labour (hr)	Machine excavating (hr)	Plant compacting (hr)	Granular material (tonne)	Cart away (cu. m)	Unit
R12.045	**Excavating trenches to receive pipes not exceeding 200 mm nominal size; grading bottoms; earthwork support; disposing of surplus excavated material by removing from site**						
R12.050	Excavating; filling in above 400 mm thick beds or coverings with granular material; not exceeding 2.0 m deep; average depth						
	500 mm	0.09	0.08	0.01	0.09	0.23	ln. m
	750 mm	0.17	0.14	0.03	0.32	0.34	ln. m
	1000 mm	0.26	0.20	0.05	0.55	0.45	ln. m
	1250 mm	0.36	0.28	0.08	0.78	0.56	ln. m
	1500 mm	0.59	0.46	0.13	1.34	0.90	ln. m
	1750 mm	0.70	0.54	0.16	1.64	1.05	ln. m
	2000 mm	1.02	0.78	0.24	2.43	1.50	ln. m
R12.055	Excavating; filling in above 400 mm thick beds or coverings with granular material; 2.0–4.0 m deep; average depth						
	2250 mm	1.23	0.95	0.28	2.73	1.69	ln. m
	2500 mm	1.38	1.07	0.32	3.19	1.88	ln. m
	2750 mm	1.53	1.18	0.35	3.57	2.06	ln. m
	3000 mm	1.68	1.29	0.39	3.95	2.25	ln. m
	3250 mm	1.83	1.40	0.43	4.33	2.44	ln. m
	3500 mm	1.98	1.52	0.47	4.71	2.63	ln. m
	3750 mm	2.13	1.63	0.50	5.09	2.81	ln. m
	4000 mm	2.28	1.74	0.54	5.47	3.00	ln. m
R12.060	Excavating; filling in above 450 mm thick beds or coverings with granular material; not exceeding 2.0 m deep; average depth						
	500 mm	0.08	0.08	0.01	0.05	0.23	ln. m
	750 mm	0.17	0.14	0.03	0.27	0.34	ln. m
	1000 mm	0.25	0.20	0.05	0.50	0.45	ln. m
	1250 mm	0.35	0.28	0.07	0.73	0.56	ln. m
	1500 mm	0.58	0.45	0.13	1.28	0.90	ln. m
	1750 mm	0.69	0.53	0.16	1.58	1.05	ln. m
	2000 mm	1.00	0.77	0.23	2.35	1.50	ln. m

	Labour (hr)	Machine excavating (hr)	Plant compacting (hr)	Granular (tonne) (tonne)	Cart away (cu. m)	Unit
R12.065						
Excavating; filling in above 450 mm thick beds or coverings with granular material; 2.0–4.0 m deep; average depth						
2250 mm	1.21	0.94	0.27	2.73	1.69	ln. m
2500 mm	1.36	1.06	0.31	3.11	1.88	ln. m
2750 mm	1.51	1.17	0.34	3.49	2.06	ln. m
3000 mm	1.66	1.28	0.38	3.87	2.25	ln. m
3250 mm	1.81	1.39	0.42	4.25	2.44	ln. m
3500 mm	1.96	1.51	0.46	4.63	2.63	ln. m
3750 mm	2.11	1.62	0.49	5.01	2.81	ln. m
4000 mm	2.26	1.73	0.53	5.39	3.00	ln. m

		Machine	Plant	Granular	Cart	
	Labour	excavating	compacting	material	away	
	(hr)	(hr)	(hr)	(tonne)	(cu. m)	Unit

R12.070	**Excavating trenches to receive pipes, 225 mm nominal size; grading bottoms; earthwork support; disposing of surplus excavated material by removing from site**						
R12.075	Excavating; filling in above 525 mm thick beds or coverings with granular material; not exceeding 2.0 m deep; average depth						
	750 mm	0.20	0.17	0.03	0.27	0.45	ln. m
	1000 mm	0.31	0.25	0.06	0.58	0.60	ln. m
	1250 mm	0.44	0.36	0.09	0.88	0.75	ln. m
	1500 mm	0.56	0.44	0.12	1.19	0.90	ln. m
	1750 mm	0.84	0.66	0.18	1.86	1.31	ln. m
	2000 mm	0.98	0.76	0.22	2.24	1.50	ln. m
R12.080	Excavating; filling in above 525 mm thick beds or coverings with granular material; 2.0–4.0 m deep; average depth						
	2250 mm	1.19	0.93	0.26	2.62	1.69	ln. m
	2500 mm	1.34	1.05	0.30	3.00	1.88	ln. m
	2750 mm	1.49	1.16	0.33	3.38	2.06	ln. m
	3000 mm	1.64	1.27	0.37	3.76	2.25	ln. m
	3250 mm	1.79	1.38	0.41	4.14	2.44	ln. m
	3500 mm	1.94	1.50	0.45	4.52	2.63	ln. m
	3750 mm	2.09	1.61	0.48	4.90	2.81	ln. m
	4000 mm	2.24	1.72	0.52	5.28	3.00	ln. m

		Labour (hr)	Machine excavating (hr)	Plant compacting (hr)	Cart away (cu. m)	Unit
R12.105	**Excavating trenches (by hand) to receive pipes not exceeding 200 mm nominal size; grading bottoms; earthwork support; disposing of surplus excavated material by removing from site**					
R12.110	Excavating; filling in above 400 mm thick beds or coverings with material arising from excavations not exceeding 2.0 m deep; average depth					
	500 mm	0.80	—	—	0.01	0.18	In. m
	750 mm	1.31	—	—	0.03	0.18	In. m
	1000 mm	1.82	—	—	0.05	0.18	In. m
	1250 mm	2.80	—	—	0.08	0.18	In. m
	1500 mm	4.54	—	—	0.13	0.24	In. m
	1750 mm	5.35	—	—	0.16	0.24	In. m
	2000 mm	7.70	—	—	0.24	0.30	In. m
R12.115	Excavating; filling in above 450 mm thick beds or coverings with material arising from excavations not exceeding 2.0 m deep; average depth					
	500 mm	0.77	—	—	0.01	0.20	In. m
	750 mm	1.29	—	—	0.03	0.20	In. m
	1000 mm	1.80	—	—	0.05	0.20	In. m
	1250 mm	2.78	—	—	0.07	0.20	In. m
	1500 mm	4.50	—	—	0.13	0.27	In. m
	1750 mm	5.31	—	—	0.16	0.27	In. m
	2000 mm	7.65	—	—	0.23	0.34	In. m

		Labour (hr)	Machine excavating (hr)	Plant compacting (hr)	Cart away (cu. m)	Unit
R12.120	**Excavating trenches to receive pipes, 225 mm nominal size; grading bottoms; earthwork support; disposing of surplus excavated material by removing from site**					
R12.125	Excavating; filling in above 525 mm thick beds or coverings with material arising from excavations not exceeding 2.0 m deep; average depth					
	750 mm	1.65	—	—	0.03	0.32	ln. m
	1000 mm	2.33	—	—	0.06	0.32	ln. m
	1250 mm	3.64	—	—	0.09	0.32	ln. m
	1500 mm	4.44	—	—	0.12	0.32	ln. m
	1750 mm	6.58	—	—	0.18	0.39	ln. m
	2000 mm	7.58	—	—	0.22	0.39	ln. m

	Labour (hr)	Machine excavating (hr)	Plant compacting (hr)	Granular material (tonne)	Cart away (cu. m)	Unit
R12.130 **Excavating trenches to receive pipes not exceeding 200 mm nominal size; grading bottoms; earthwork support; disposing of surplus excavated material by removing from site**						
R12.135 Excavating; filling in above 400 mm thick beds or coverings with granular material; not exceeding 2.0 m deep; average depth						
500 mm	0.83	—	0.01	0.09	0.23	ln. m
750 mm	1.42	—	0.03	0.32	0.34	ln. m
1000 mm	2.01	—	0.05	0.55	0.45	ln. m
1250 mm	3.07	—	0.08	0.78	0.56	ln. m
1500 mm	5.00	—	0.13	1.34	0.90	ln. m
1750 mm	5.92	—	0.16	1.64	1.05	ln. m
2000 mm	8.54	—	0.24	2.43	1.50	ln. m
R12.140 Excavating; filling in above 450 mm thick beds or coverings with granular material; not exceeding 2.0 m deep; average depth; 500 mm	0.79	—	0.01	0.05	0.23	ln. m
750 mm	1.38	—	0.03	0.27	0.34	ln. m
1000 mm	1.97	—	0.05	0.50	0.45	ln. m
1250 mm	3.03	—	0.07	0.73	0.56	ln. m
1500 mm	4.95	—	0.13	1.28	0.90	ln. m
1750 mm	5.86	—	0.16	1.58	1.05	ln. m
2000 mm	8.46	—	0.23	2.35	1.50	ln. m

	Labour (hr)	Machine excavating (hr)	Plant compacting (hr)	Granular material (tonne)	Cart away (cu. m)	Unit
R12.145	**Excavating trenches to receive pipes, 225 mm nominal size; grading bottoms; earthwork support; disposing of surplus excavated material by removing from site**					
R12.150	Excavating; filling in above 525 mm thick beds or coverings with granular material; not exceeding 2.0 m deep; average depth					
	750 mm 1.74	—	0.03	0.27	0.45	ln. m
	1000 mm 2.53	—	0.06	0.58	0.60	ln. m
	1250 mm 3.94	—	0.09	0.88	0.75	ln. m
	1500 mm 4.85	—	0.12	1.19	0.90	ln. m
	1750 mm 7.22	—	0.18	1.86	1.31	ln. m
	2000 mm 8.36	—	0.22	2.24	1.50	ln. m

	Labour (hr)				Granular material (tonne)	Unit
R12.205	**Beds and coverings; granular material, 10 mm nominal size pea shingle, to be obtained off site**					
R12.210	50 mm bed					
	450 mm wide	0.07	—	—	—	0.038	ln. m
	600 mm wide	0.10	—	—	—	0.050	ln. m
	750 mm wide	0.12	—	—	—	0.063	ln. m
R12.215	100 mm bed					
	450 mm wide	0.15	—	—	—	0.075	ln. m
	600 mm wide	0.20	—	—	—	0.100	ln. m
	750 mm wide	0.24	—	—	—	0.125	ln. m
R12.220	150 mm bed					
	450 mm wide	0.22	—	—	—	0.113	ln. m
	600 mm wide	0.29	—	—	—	0.150	ln. m
	750 mm wide	0.37	—	—	—	0.188	ln. m
R12.225	100 mm bed and filling to half height of pipe					
	450 mm wide to 100 mm pipe	0.22	—	—	—	0.111	ln. m
	600 mm wide to 100 mm pipe	0.30	—	—	—	0.152	ln. m
	750 mm wide to 100 mm pipe	0.37	—	—	—	0.192	ln. m
	450 mm wide to 150 mm pipe	0.24	—	—	—	0.121	ln. m
	600 mm wide to 150 mm pipe	0.33	—	—	—	0.169	ln. m
	750 mm wide to 150 mm pipe	0.42	—	—	—	0.216	ln. m
	600 mm wide to 225 mm pipe	0.37	—	—	—	0.189	ln. m
	750 mm wide to 225 mm pipe	0.48	—	—	—	0.248	ln. m
	600 mm wide to 300 mm pipe	0.38	—	—	—	0.196	ln. m
	750 mm wide to 300 mm pipe	0.52	—	—	—	0.268	ln. m

		Labour (hr)				Granular material (tonne)	Unit
R12.230	150 mm bed and filling to half height of pipe						
	450 mm wide to 100 mm pipe	0.29	—	—	—	0.149	ln. m
	600 mm wide to 100 mm pipe	0.39	—	—	—	0.202	ln. m
	750 mm wide to 100 mm pipe	0.50	—	—	—	0.255	ln. m
	450 mm wide to 150 mm pipe	0.31	—	—	—	0.159	ln. m
	600 mm wide to 150 mm pipe	0.43	—	—	—	0.219	ln. m
	750 mm wide to 150 mm pipe	0.54	—	—	—	0.279	ln. m
	600 mm wide to 225 mm pipe	0.47	—	—	—	0.239	ln. m
	750 mm wide to 225 mm pipe	0.61	—	—	—	0.311	ln. m
	600 mm wide to 300 mm pipe	0.48	—	—	—	0.246	ln. m
	750 mm wide to 300 mm pipe	0.65	—	—	—	0.331	ln. m
R12.235	Bed and covering						
	450 mm wide × 300 mm thick to 100 mm pipe	0.44	—	—	—	0.206	ln. m
	450 mm wide × 400 mm thick to 100 mm pipe	0.55	—	—	—	0.281	ln. m
	600 mm wide × 300 mm thick to 100 mm pipe	0.55	—	—	—	0.281	ln. m
	600 mm wide × 400 mm thick to 100 mm pipe	0.74	—	—	—	0.382	ln. m
	750 mm wide × 300 mm thick to 100 mm pipe	0.70	—	—	—	0.357	ln. m
	750 mm wide × 400 mm thick to 100 mm pipe	0.94	—	—	—	0.482	ln. m
	450 mm wide × 350 mm thick to 150 mm pipe	0.43	—	—	—	0.222	ln. m
	450 mm wide × 450 mm thick to 150 mm pipe	0.58	—	—	—	0.297	ln. m
	600 mm wide × 350 mm thick to 150 mm pipe	0.60	—	—	—	0.310	ln. m
	600 mm wide × 450 mm thick to 150 mm pipe	0.80	—	—	—	0.410	ln. m
	750 mm wide × 350 mm thick to 150 mm pipe	0.77	—	—	—	0.397	ln. m
	750 mm wide × 450 mm thick to 150 mm pipe	1.02	—	—	—	0.523	ln. m
	600 mm wide × 450 mm thick to 225 mm pipe	0.69	—	—	—	0.351	ln. m
	600 mm wide × 600 mm thick to 225 mm pipe	0.98	—	—	—	0.502	ln. m
	750 mm wide × 450 mm thick to 225 mm pipe	0.91	—	—	—	0.464	ln. m
	750 mm wide × 600 mm thick to 225 mm pipe	1.27	—	—	—	0.652	ln. m
	600 mm wide × 500 mm thick to 300 mm pipe	0.62	—	—	—	0.317	ln. m
	600 mm wide × 600 mm thick to 300 mm pipe	0.81	—	—	—	0.417	ln. m
	750 mm wide × 500 mm thick to 300 mm pipe	0.86	—	—	—	0.442	ln. m
	750 mm wide × 600 mm thick to 300 mm pipe	1.11	—	—	—	0.568	ln. m

	5/3.5 Mixer labour (hr)	7/5 Mixer labour (hr)	10/7 Mixer labour (hr)		Concrete (cu. m)	Unit
R12.305	**Beds and coverings; plain in situ concrete**					
R12.310	100 mm bed					
450 mm wide	0.31	0.25	0.22	—	0.045	ln. m
600 mm wide	0.41	0.33	0.29	—	0.060	ln. m
750 mm wide	0.51	0.41	0.36	—	0.075	ln. m
R12.315	150 mm bed					
450 mm wide	0.46	0.37	0.32	—	0.068	ln. m
600 mm wide	0.61	0.50	0.43	—	0.090	ln. m
750 mm wide	0.77	0.62	0.54	—	0.113	ln. m
R12.320	100 mm bed and filling to half height of pipe					
450 mm wide to 100 mm pipe	0.45	0.36	0.32	—	0.066	ln. m
600 mm wide to 100 mm pipe	0.62	0.50	0.44	—	0.091	ln. m
750 mm wide to 100 mm pipe	0.78	0.63	0.55	—	0.115	ln. m
450 mm wide to 150 mm pipe	0.49	0.40	0.35	—	0.072	ln. m
600 mm wide to 150 mm pipe	0.69	0.56	0.48	—	0.101	ln. m
750 mm wide to 150 mm pipe	0.88	0.71	0.62	—	0.129	ln. m
600 mm wide to 225 mm pipe	0.77	0.62	0.54	—	0.113	ln. m
750 mm wide to 225 mm pipe	1.01	0.81	0.71	—	0.148	ln. m
600 mm wide to 300 mm pipe	0.80	0.64	0.56	—	0.117	ln. m
750 mm wide to 300 mm pipe	1.09	0.89	0.77	—	0.161	ln. m
R12.325	150 mm bed and filling to half height of pipe					
450 mm wide to 100 mm pipe	0.61	0.49	0.43	—	0.089	ln. m
600 mm wide to 100 mm pipe	0.82	0.67	0.58	—	0.121	ln. m
750 mm wide to 100 mm pipe	1.04	0.84	0.73	—	0.153	ln. m
450 mm wide to 150 mm pipe	0.65	0.52	0.46	—	0.095	ln. m
600 mm wide to 150 mm pipe	0.89	0.72	0.63	—	0.131	ln. m
750 mm wide to 150 mm pipe	1.14	0.92	0.80	—	0.167	ln. m
600 mm wide to 225 mm pipe	0.97	0.79	0.69	—	0.143	ln. m
750 mm wide to 225 mm pipe	1.26	1.02	0.89	—	0.186	ln. m
600 mm wide to 300 mm pipe	1.00	0.81	0.71	—	0.147	ln. m
750 mm wide to 300 mm pipe	1.35	1.09	0.95	—	0.198	ln. m
R12.330	Bed and covering					
450 mm wide × 400 mm thick to 100 mm pipe	1.02	0.80	0.69	—	0.168	ln. m
450 mm wide × 450 mm thick to 150 mm pipe	1.08	0.85	0.73	—	0.178	ln. m
525 mm wide × 525 mm thick to 225 mm pipe	1.31	1.03	0.88	—	0.216	ln. m
600 mm wide × 600 mm thick to 300 mm pipe	1.52	1.19	1.02	—	0.250	ln. m

		Labour (hr)	Pipe/ fittings (ln. m)	Couplings (no.)	Lubricant (kg)	Concrete (cu. m)	Unit
R12.405	**Pipes; vitrified clay pipes and fittings; SuperSleve: push-fit polypropylene flexible couplings**						
R12.410	100 mm pipework in trenches						
	laid in position	0.21	1.05	0.656	0.016	—	ln. m
	laid in position, in runs not exceeding 3.0 m	0.28	1.05	0.656	0.016	—	ln. m
	laid in position, vertical	0.31	1.05	0.656	0.016	—	ln. m
	bends; Ref. 8-20	0.25	1.05	1.050	0.026	—	no.
	rest bends; Ref. 19	0.25	1.05	1.050	0.026	—	no.
	junction, 100 × 100 mm; Ref. 22	0.22	1.05	2.100	0.053	—	no.
	adapters Ref. AD	0.05	1.05	—	0.026	—	no.
R12.415	150 mm pipework in trenches						
	laid in position	0.29	1.05	0.600	0.045	—	ln. m
	laid in position, in runs not exceeding 3.0 m	0.38	1.05	0.600	0.045	—	ln. m
	laid in position, vertical	0.46	1.05	0.600	0.045	—	ln. m
	bends; Ref. 8-20	0.33	1.05	1.050	0.079	—	no.
	rest bends; Ref. 19	0.33	1.05	1.050	0.079	—	no.
	junction, 150 × 150 mm; Ref. 22 junction, 150 × 100 mm; Ref. 22 * (1/100 and 1/150 mm coupling)	0.29	1.05	*1.050	0.065	—	no.
	Taper 150–100 mm; Ref. 46 * (1/100 mm coupling)	0.33	1.05	*1.050	0.105	—	no.
R12.420	**Vitrified clay accessories**						
R12.425	Gullies; HepSleve; joint to pipe; bedding and surrounding in concrete 1:3:6 nominal mix, 20 mm aggregate						
	100 mm outlet, Ref. 148A, low-back P trap; hopper, Ref. 261, square with integral back inlet	1.17	1.05	1.05	0.026	0.050	no.
	100 mm outlet, Ref. 191A, trapped; plastic stopper, square 150 × 150 integral vertical back inlet	1.00	1.05	—	—	0.06	no.
	100 mm outlet, Ref. 196, trapped, square 200 × 200 × 485 mm deep; rodding eye stopper	1.31	1.05	—	—	0.07	no.
	galvanized bucket, Ref. 196A	0.05	1.05	—	—	—	no.

		Labour (hr)	Pipe/ fittings (ln. m)	Couplings (no.)	Lubricant (kg)	Concrete (cu. m)	Unit
R12.430	Inspection chamber, supersleve; joint to pipe 100 mm outlet, Ref. IC.69 base, straight through; ICC coupling; IC.200 raising piece with integral alloy cover and frame	1.55	1.05	1.05	0.007	—	no.
	100 mm outlet, Ref. IC.63 or IC.65 base, left- or right-hand junction; ICC coupling; IC.200 raising piece with integral alloy cover and frame	1.55	1.05	2.10	0.013	—	no.
	100 mm outlet, Ref. IC.64 base, left- and right-hand junction; ICC coupling; ICC.200 raising piece with integral alloy cover and frame	1.55	1.05	3.15	0.020	—	no.
R12.435	**Aluminium alloy accessories**						
R12.440	Rodding points; SuperSleve; joint to pipe; surrounding in concrete; 20 mm aggregate						
	100 mm outlet, Ref. 262; 205 × 150 mm	0.64	1.05	—	—	0.05	no.
R12.445	Gratings; placing in position						
	Ref. 1002; 150 × 150 mm	0.05	1.05	—	—	—	no.
	Ref. 1002; 225 × 225 mm	0.05	1.05	—	—	—	no.

	Labour (hr)	Pipe/ fittings (ln. m)	Couplings (no.)	Lubricant (kg)	Concrete (cu. m)	Unit
R12.450	**Pipes; vitrified clay pipes and fittings, HepSeal; spigot and socket joints with rubber sealing rings**					
R12.455	150 mm pipework in trenches					
	laid in position	0.33	1.05	—	0.010	—	ln. m
	laid in position, in runs not exceeding 3.0 m	0.44	1.05	—	0.010	—	ln. m
	bends Ref. 8,14	0.44	1.05	—	0.007	—	no.
	rest bends Ref. 19	0.44	1.05	—	0.007	—	no.
	junction, 150×150 mm; Ref. 22	0.29	1.05	—	0.013	—	no.
	junction, 150×100 mm; Ref. 22A	0.29	1.05	1.05	0.013	—	no.
R12.460	225 mm pipework in trenches					
	laid in position	0.38	1.05	—	0.015	—	ln. m
	laid in position, in runs not exceeding 3.0 m	0.51	1.05	—	0.015	—	ln. m
	bends Refs. 8,14,17,18	0.50	1.05	—	0.009	—	no.
	rest bend Ref. 19	0.50	1.05	—	0.009	—	no.
	junction, 225×225 mm; Ref. 22	0.40	1.05	—	0.017	—	no.
	junction, 225×150 mm; Ref. 22A	0.40	1.05	1.05	0.017	—	no.
	junction, 225×100 mm; Ref. 22A	0.40	1.05	1.05	0.017	—	no.
R12.465	300 mm pipework in trenches					
	laid in position	0.53	1.05	—	0.014	—	ln. m
	laid in position, in runs not exceeding 3.0 m	0.73	1.05	—	0.014	—	ln. m
	bends Refs. 8,14,17,18	0.73	1.05	—	0.006	—	no.
	rest bend Ref. 19	0.73	1.05	—	0.006	—	no.
	junction, 300×300 mm; Ref. 22	0.67	1.05	—	0.011	—	no.
	junction, 300×225 mm; Ref. 22A	0.67	1.05	1.05	0.011	—	no.
	junction, 300×150 mm; Ref. 22A	0.67	1.05	1.05	0.011	—	no.
	junction, 300×100 mm; Ref. 22A	0.67	1.05	1.05	0.011	—	no.

		Labour (hr)	Fittings (no.)	Concrete (cu. m)	Engineering bricks (no.)	Cement mortar (cu. m)	Unit
R12.475	**Concrete accessories**						
R12.480	Road gullies; BS.556; cement mortar joint to pipe; bedding and surrounding in concrete						
	375 mm diameter × 750 mm deep; 150 mm outlet; trapped	4.95	1.025	0.55	—	—	no.
	two courses of engineering bricks, class B, in cement mortar, in raising pieces, to three sides of gulley	1.22	—	—	27.00	0.016	no.
R12.485	**Ductile iron accessories**						
R12.490	Road gratings and frames; BS.497 grade B; bedding frame in cement mortar; haunching frame with cement mortar;						
	Ref. GB-325; 3-sided	0.22	1.025	—	—	0.019	no.
	Ref. GB-325; 4-sided	0.25	1.025	—	—	0.020	no.

		Labour (hr)	Machine excavating (hr)				Unit
R12.505	**Manholes**						
R12.510	Excavating pits (by machine) to receive bases of manholes						
	maximum depth not exceeding						
	1.00 m	0.22	0.22	—	—	—	cu. m
	2.00 m	0.22	0.22	—	—	—	cu. m
	4.00 m	0.27	0.27	—	—	—	cu. m
R12.515	Excavating pits (by hand) to receive bases of manholes						
	maximum depth not exceeding						
	1.00 m	3.30	—	—	—	—	cu. m
	2.00 m	4.13	—	—	—	—	cu. m
	4.00 m	6.60	—	—	—	—	cu. m
R12.520	Levelling and compacting bottoms of excavations in manholes	0.08	—	—	—	—	sq. m

		5/3.5 Mixer labour (hr)	10/7 Mixer labour (hr)	Concrete (cu. m)	Cement mortar (cu. m)	Steel fabric (sq. m)	Unit
R12.525	Plain in situ concrete beds in manhole bases						
	100–150 mm thick	6.50	4.51	1.00	—	—	cu. m
	150–300 mm thick	6.08	4.09	1.00	—	—	cu. m
R12.530	Plain in situ concrete surrounds to manholes						
	100–150 mm thick	7.85	5.86	1.00	—	—	cu. m
	150–300 mm thick	7.45	5.46	1.00	—	—	cu. m
R12.535	Plain in situ concrete benching in bottoms of manholes; surface finished with 25 mm cement mortar trowelled smooth; average						
	225 mm thick	2.96	—	0.23	0.025	—	sq. m
	300 mm thick	3.76	—	0.30	0.025	—	sq. m
	450 mm thick	5.35	—	0.45	0.025	—	sq. m
R12.540	Reinforced in situ concrete in suspended slabs to manholes						
	100–150 mm thick	7.57	5.58	1.00	—	—	cu. m
	150–300 mm thick	6.86	4.87	1.00	—	—	cu. m
R12.545	Steel fabric reinforcement, with one width side laps and one width end laps in suspended slabs to manholes, reference;						
	A 142, 200 mm side lap, 200 mm end lap	0.20	—	—	—	1.220	sq. m
	A 252, 200 mm side lap, 200 mm end lap	0.25	—	—	—	1.220	sq. m
	B 196, 100 mm side lap, 200 mm end lap	0.25	—	—	—	1.170	sq. m
R12.550	Circular cutting fabric						
	A 142	0.21	—	—	—	0.056	ln. m
	A 252	0.31	—	—	—	0.050	ln. m
	B 196	0.31	—	—	—	0.053	ln. m

		Labour (hr)			Chamber units (no.)	Mortar (cu. m)	Unit
R12.605	**Manholes; precast concrete inspection chamber units; Mono; rebated joints; bedded and pointed with cement mortar**						
R12.610	Chamber units						
	Type A; 610 × 455 mm internal size						
	100 mm deep	0.50	—	—	1.05	0.001	no.
	150 mm deep	0.65	—	—	1.05	0.001	no.
	250 mm deep	0.75	—	—	1.05	0.001	no.
	Type B; 760 × 610 mm internal size						
	150 mm deep	0.75	—	—	1.05	0.002	no.
	Type C; 990 × 610 mm internal size;						
	150 mm deep	0.80	—	—	1.05	0.002	no.
R12.615	Reducer slabs						
	Type B-A; 760 × 610 to 610 × 455 mm						
	128 mm deep	0.75	—	—	1.05	0.001	no.
	Type C-B; 990 × 610 to 760 × 610 mm						
	128 mm deep	0.50	—	—	1.05	0.001	no.
	Type C-A; 990 × 610 to 610 × 455 mm						
	128 mm deep	0.75	—	—	1.05	0.001	no.
R12.620	Base shutter units						
	Type A; 610 × 455 mm internal size						
	380 mm deep	0.75	—	—	1.05	0.001	no.
	Type B; 760 × 610 mm internal size						
	380 mm deep	1.20	—	—	1.05	0.002	no.
	Type C; 990 × 610 mm internal size						
	380 mm deep	1.35	—	—	1.05	0.002	no.
R12.625	Covers						
	regular cover	0.33	—	—	1.05	—	no.
	super cover	0.33	—	—	1.05	—	no.
	utility cover	0.33	—	—	1.05	—	no.
	rodding eye cover	0.17	—	—	1.05	—	no.
R12.630	Frames						
	standard frame	0.33	—	—	1.05	0.001	no.
	frame for concrete inspection cover and						
	frame	0.33	—	—	1.05	0.001	no.
	utility frame	0.33	—	—	1.05	0.001	no.
	rodding eye frame	0.50	—	—	1.05	0.001	no.

		Labour (hr)		Machine excavating (hr)	Chamber rings (no.)	Cement mortar (cu. m)	Unit
R12.635	**Manholes; precast concrete manhole chamber units; BS.556; rebated joints; bedding and jointing in cement mortar**						
R12.640	Shaft rings						
	675 mm diameter						
	300 mm deep	0.58	—	0.58	1.05	0.001	no.
	460 mm deep	0.67	—	0.67	1.05	0.001	no.
	610 mm deep	0.81	—	0.81	1.05	0.001	no.
	760 mm deep	0.87	—	0.87	1.05	0.001	no.
	910 mm deep	0.93	—	0.93	1.05	0.001	no.
R12.645	Chamber rings						
	1050 mm diameter						
	300 mm deep	0.81	—	0.81	1.05	0.002	no.
	460 mm deep	0.87	—	0.87	1.05	0.002	no.
	610 mm deep	0.93	—	0.93	1.05	0.002	no.
	760 mm deep	0.99	—	0.99	1.05	0.002	no.
	910 mm deep	1.04	—	1.04	1.05	0.002	no.
	1200 mm diameter						
	300 mm deep	0.99	—	0.99	1.05	0.003	no.
	460 mm deep	1.04	—	1.04	1.05	0.003	no.
	610 mm deep	1.16	—	1.16	1.05	0.003	no.
	760 mm deep	1.22	—	1.22	1.05	0.003	no.
	910 mm deep	1.28	—	1.28	1.05	0.003	no.
R12.650	Taper sections						
	1050 to 675 mm diameter	1.04	—	1.04	1.05	0.002	no.
	1200 to 675 mm diameter	1.04	—	1.04	1.05	0.003	no.
R12.655	Cover slabs						
	675 mm diameter						
	light duty	0.50	—	0.50	1.05	0.001	no.
	heavy duty	0.58	—	0.58	1.05	0.001	no.
	1050 mm diameter						
	light duty	0.58	—	0.58	1.05	0.002	no.
	heavy duty	0.67	—	0.67	1.05	0.002	no.
	1200 mm diameter						
	light duty	0.60	—	0.60	1.05	0.003	no.
	heavy duty	0.69	—	0.69	1.05	0.003	no.

		Labour (hr)	Bricks (no.)	Mortar		Unit
				No frog (cu. m)	One frog (cu. m)	
R12.705	**Manholes; engineering bricks, 215× 103 ×65 mm, Class B; in cement mortar**					
R12.710	Walls in manholes					
	215 mm thick	3.39	—	120.87	0.045	0.074	sq. m
R12.715	Extra over engineering bricks, in any bond in any mortar for facework; walls or the like					
	stretcher bond	0.46	—	—	—	—	sq. m
	English bond	0.50	—	—	—	—	sq. m
R12.720	Building in ends of pipes; 215 mm brickwork					
	100 mm pipe	0.27	—	—	—	—	no.
	150 mm pipe	0.33	—	—	—	—	no.
	225 mm pipe	0.37	—	—	—	—	no.
	300 mm pipe	0.40	—	—	—	—	no.
R12.725	**Manholes; mortar, cement and sand**					
R12.730	12 mm work to walls on brickwork base; steel trowelled					
	over 300 mm wide	0.67	—	—	0.012	0.012	sq. m
	not exceeding 300 mm wide	1.34	—	—	0.012	0.012	sq. m

	Labour (hr)	Metalwork (no.)	Mortar (cu. m)	Concrete (cu. m)	Unit
R12.805 **Manholes; metalwork**					
R12.810 Step irons; cast iron; galvanized; general purpose pattern; building into brickwork					
115 mm tail	0.08	—	1.05	—	—	no.
230 mm tail	0.12	—	1.05	—	—	no.
R12.815 Manhole covers and frames; coated; bedding frame in cement mortar; bedding cover in grease and sand					
Grade A cover and frame					
Ref. MA-50	1.33	—	1.00	0.015	—	no.
Ref. MA-60	1.67	—	1.00	0.012	—	no.
Grade B Class 1 cover and frame					
Ref. MB1-50	0.83	—	1.00	0.017	—	no.
Ref. MB1-60	1.00	—	1.00	0.011	—	no.
Grade B Class 2 cover and frame					
Ref. MB2-50	0.83	—	1.00	0.008	—	no.
Ref. MB2-60	1.00	—	1.00	0.008	—	no.
Ref. MB2-60/45	1.00	—	1.00	0.012	—	no.
Ref. MB2-60/60	1.33	—	1.00	0.015	—	no.
R12.820 Inspection chamber covers and frames; bedding frame in cement mortar					
Grade C Single seal					
Ref. MC1-60/45	0.58	—	1.00	0.004	—	no.
Ref. MC1-60/60	0.67	—	1.00	0.005	—	no.
Grade C Double seal					
Ref. MC2-60/45	0.58	—	1.00	0.005	—	no.
Ref. MC2-60/60	0.67	—	1.00	0.007	—	no.
Grade B Single seal; recessed					
Ref. DC5293	1.00	—	1.00	0.009	—	no.
Ref. DC5294	1.33	—	1.00	0.012	—	no.
Grade C Double seal; recessed					
Ref. MC2R-60/45	1.00	—	1.00	0.006	—	no.
Ref. MC2R-60/60	1.33	—	1.00	0.008	—	no.
R12.825 Broadstel Universal access covers and frames, with rubber seals and locking devices; bedding frames in cement mortar;					
Medium duty cover and frame					
600 × 450 mm (DC.349B)	0.83	—	1.00	0.010	—	no.
600 × 600 mm (DC.349C)	1.25	—	1.00	0.012	—	no.
Medium heavy duty cover and frame					
600 × 450 mm (DC.350B)	0.92	—	1.00	0.010	—	no.

	Labour (hr)	Straight channel (ln. m)	Channel bends (no.)	Channel fittings (no.)	Cement mortar (cu. m)	Unit
R12.905 **Manholes; channels in bottoms; vitrified clay; cement mortar joints; bedding in cement mortar**						
R12.910 Half section; straight						
100 mm; effective length						
600 mm	0.23	0.63	—	—	0.003	no.
750 mm	0.27	0.79	—	—	0.004	no.
900 mm	0.30	0.95	—	—	0.005	no.
1050 mm	0.33	1.10	—	—	0.006	no.
1200 mm	0.42	1.26	—	—	0.007	no.
150 mm; effective length						
600 mm	0.40	0.63	—	—	0.007	no.
750 mm	0.43	0.79	—	—	0.008	no.
900 mm	0.47	0.95	—	—	0.010	no.
1050 mm	0.50	1.10	—	—	0.012	no.
1200 mm	0.58	1.26	—	—	0.013	no.
225 mm; effective length						
600 mm	0.65	0.63	—	—	0.015	no.
750 mm	0.68	0.79	—	—	0.019	no.
900 mm	0.72	0.95	—	—	0.023	no.
1050 mm	0.75	1.10	—	—	0.026	no.
1200 mm	0.83	1.26	—	—	0.030	no.
300 mm; effective length						
600 mm	0.82	0.63	—	—	0.030	no.
750 mm	0.85	0.79	—	—	0.038	no.
900 mm	0.88	0.95	—	—	0.045	no.
1050 mm	0.92	1.10	—	—	0.053	no.
1200 mm	1.08	1.26	—	—	0.060	no.

	Labour (hr)	Straight channel (ln. m)	Channel bends (no.)	Channel fittings (no.)	Cement mortar (cu. m)	Unit
R12.915 Half section; curved						
100 mm; effective length						
600 mm	0.32	—	4.00	—	0.003	no.
750 mm	0.35	—	5.00	—	0.004	no.
900 mm	0.38	—	6.00	—	0.005	no.
1200 mm	0.50	—	8.00	—	0.007	no.
150 mm; effective length						
600 mm	0.40	—	3.00	—	0.007	no.
750 mm	0.43	—	4.00	—	0.008	no.
900 mm	0.47	—	5.00	—	0.010	no.
1200 mm	0.58	—	7.00	—	0.013	no.
225 mm; effective length						
600 mm	0.73	—	3.00	—	0.015	no.
750 mm	0.77	—	4.00	—	0.019	no.
900 mm	0.80	—	4.00	—	0.023	no.
1200 mm	0.92	—	6.00	—	0.030	no.
300 mm; effective length						
600 mm	0.90	—	3.00	—	0.030	no.
750 mm	0.93	—	3.00	—	0.038	no.
900 mm	0.97	—	4.00	—	0.045	no.
1200 mm	1.17	—	5.00	—	0.060	no.
R12.920 Half section; channel junction						
100 mm	0.37	—	—	1.05	0.003	no.
150 mm	0.50	—	—	1.05	0.003	no.
225 mm	0.75	—	—	1.05	0.004	no.
300 mm	1.00	—	—	1.05	0.005	no.
R12.925 Half section; tapered						
100–150 mm	0.50	—	—	1.05	0.003	no.
150–225 mm	0.75	—	—	1.05	0.009	no.
225–300 mm	1.00	—	—	1.05	0.022	no.
R12.930 Half section; branch bends						
100 mm	0.37	—	—	1.05	0.009	no.
150 mm	0.50	—	—	1.05	0.017	no.
225 mm	0.75	—	—	1.05	0.020	no.
300 mm	1.00	—	—	1.05	0.024	no.
R12.935 Three quarter section; branch bends						
100 mm	0.37	—	—	1.05	0.003	no.
150 mm	0.50	—	—	1.05	0.006	no.
225 mm	0.75	—	—	1.05	0.015	no.
300 mm	1.00	—	—	1.05	0.036	no.